Appomattox River Seay Stories

"And I took my old boat out the canal over yonder and used it for a ferry and made a lot of money . . . round trip 5 cents."

COVER: *Very few photographs of Virginia batteaux have survived, including this remarkable post-card view of Jim Seay's boat. Around 1902 he moved it into the Appomattox River at Matoaca, added a cabin, and for eleven years carried passengers between Matoaca and Ferndale Park until he was replaced by a bridge. On the Ferndale side the ferry landed at a long wooden dock, as pictured on the back cover. This postcard, entitled "Ferry Connecting Petersburg with Matoaca, Petersburg, Va.," was published by S.S. Kresge, Detroit, Michigan, about 1910. Courtesy of Lee A. Wallace, Jr., and Fred R. Bell.*

OPPOSITE PAGE: *Drawing of S. D. Morton's wax seal discovered on a letter dated Petersburg, Va., May 31, 1826, in the Virginia State Library's collecton of Upper Appomattox Company correspondence. Morton was the company's "clerk," or secretary.*

Appomattox River Seay Stories

*Reminiscences of James Washington Seay
The Last of the Appomattox River
Batteaumen*

Edited by William E. Trout, III, PhD
With the Assistance of R. Dulaney Ward, Jr.

Published by
The Historic Petersburg Foundation
& The Virginia Canals and Navigations Society

1992

Copyright 1992 by the Historic Petersburg Foundation, Inc.

All rights reserved. No part of this book may be used without permission of the publisher.

Published by the Historic Petersburg Foundation, P. O. Box 691, Petersburg, VA 23804, and by the Virginia Canals and Navigations Society, c/o the Alexandria Waterfront Museum, 44 Canal Center, Alexandria, VA 22314

International Standard Book Number 1-882365-00-3

Library of Congress Catalog Card Number: 92-073975

Printed in the United States of America by The Dietz Press, Richmond, Virginia

First Printing, November 1992

Designed and set in type by Dulaney Ward. Set in Palatino using Microsoft Word and Aldus Pagemaker for Windows, and an Apple Laserwriter IINT. Images scanned using a Hewlett-Packard Scanjet.

"A wonderful thing . . . . If you wanted to take down some great man's voice, you'd still have it."
Jim Seay, on tape recorders

JIM SEAY THINKS BACK TO WHEN HE BOUGHT HIS WATCH

(In an interview with Dick Pitts)

Dick Pitts: "Let me see that watch. I'm a watch man, you know."

Jim Seay: "I bought that watch in 1900. Elgin. Fell in the river once and never run since.... Paid ten dollars for it. About 19 hundred and —"

P: "What'd you say if I started it up?"

S: "Ain't been moving in 10 years."

P: (winding it up) "Well why is it running now?"

S: "I don't know. Ain't moving in—let me see. Let me see now. I bought it in 19— .... I was running my ferry boat and ran the ferry boat from 1900 to 19 hundred 11. I bought it somewhere along 19 hundred and 8. Somewhere along there, and about 10 or 15 years ago I fell in the river with that in my pocket and it stopped running and I never fooled with it anymore."

P: "O.K., it's 9 minutes after 12. I'm going to set it for you ..."

S: "I'll be doggoned ... It's running! ... You *must* be a watch man!"

Photo courtesy of Mr. & Mrs. Rubin Traylor, with thanks to Danny Rubin Partin

In Memory of Jim Seay, 1878-1973, on the centennial of his debut as a batteauman

"Had a sweep on the boat, which is a big paddle."

This is artist John Exley's drawing of the only known surviving batteau steering oar, saved by Jim Seay and now on display in Reynolds Metals Tidewater Connection Locks Park at 12th and Byrd streets in Richmond. Seay recalled that the oar was made by Peter George Hunter, a black man, who lived in a log cabin by the river five miles above Matoaca on the Chesterfield County side. Mr. Hunter worked on boats all his life, loading and operating them under Jim Seay and a Mr. Peter Andrews. Mr. Hunter painted his initials, "PGH," on the blade of the steering oar, but he could neither read nor write and got his letters a little mixed up. The 16-foot oar swiveled on a metal fork attached to the stern of the batteau.

CONTENTS

Illustrations	xi
Preface	xiii
Introduction	xvii
A Biography of James Washington Seay, by Gary Dalton	1

Chapter 1: Jim Seay, Batteauman

How Jim Seay Poled His Batteau	15
He Discovers an Early Batteau Pole	16
How Aunt Harriet's Furniture Came to Matoaca	18
Mr. Bowman and the Barrel of Whiskey	19
The Boatman's Lost Gold	20
The Batteau Built Too Big for the Locks	21
The Canal Basin in Petersburg, and the Mills There	22
How the Canal Aqueduct Fell Down	24

Chapter 2: Jim Seay, Ferryman

Jim Seay's Batteau Becomes a Ferry Boat at Matoaca	31
How He Laid Out His Ferry Cable	32
He Invents a Perpetual Motion Batteau Ferry	33

Chapter 3: Jim Seay, Outdoorsman

On the Outdoor Life	35
Jim Seay Goes Fishing with Mary Louise	36
Escaping the Ice Jam	37
Black's Canal and Mr. Webber's Carved Stone	38
Mr. Webber's Stone (Part 2)	38
The Strange Rocks from the Pothole	39
Jim Seay Finds the Body	40
Jim Seay's Lost Log Canoe	42
How the Historical Society Got Its Dugout Canoe	43

Chapter 4: Jim Seay's Houseboats

A Dam Which Was Never Begun, and a Mansion Which Was Never Finished	45
How Jim Seay Ran His Level	46
Jim Seay's Houseboat	49
How the Preacher's Son Fell in the River	49

Chapter 5: Ferndale Park

Ferndale Park in Its Heyday, and What Happened to the Little Dog	51
Ferndale Park (Part 2)	53
The Day the Canal Blew Out at Ferndale Park	57
How VEPCO Widened the Canal with Steam Drills	58

Chapter 6: Lake Chesdin

How the Bed of Lake Chesdin Was Cleared Out	61
How Colonel Rose Got His Tail Wet	63

Chapter 7: Jim Seay at Work

Blasting the Mystery Hole in Lafayette's Rock	65
Lafayette's Rock (Part 2)	66
How Jim Seay Built Lake Margaret Dam	67
He Catches a Fish and Saves a Snake at Lake Margaret	69
How Jim Seay Rebuilt the Matoaca Mills Dam	70
The Car That Came Out of the Canal	71
The Escape Tunnel at Olive Hill Plantation	73
Jim Seay Re-discovers a Lost Art at Olive Hill	74
Why He Had to Swim across Cohoon Pond	75
Old Syl Belcher and His Watermelon	75

Chapter 8: Jim Seay, Historian

Beadle's Dummy	77
Stingy Black Billy	78
How Bosh Regan Lost His Arm	78
How the Trunk Factory Made Its Move	81
Millions for Defence, but Not One Cent for Tribute	82
Why Dick Jones Went to Jail after He Died	83
A Tour Guide to Petersburg's Canals	85
For Further Reading	89
Index	93

ILLUSTRATIONS

Jim Seay's Batteau Ferry Boat	cover
The Upper Appomattox Company's Handwritten Seal	iii
Jim Seay, 1878-1973	vii
Peter George Hunter's Batteau Steering Oar	viii
Jim Seay at Home	xvi
Jim Seay with his Gas-powered Saw	2
An Original Batteau from Richmond's Great Basin	4
Jim Seay's Home, Which He Moved to Matoaca	6
Low Wall Sluice for Batteaux	8
An Upper Appomattox Company Stock Certificate	9
Ferndale Park's First "Batteau Day"	10
Launching the *Lord Chesterfield* at Farmville	11
Caudle's Lock as Jim Seay Remembered It	12
Jim Seay with His Model of Caudle's Lock	13
Close-up of the Lock Model	13
Lewis Miller's 1853 View of the Appomattox	14
An Exciting Moment for a Batteau	16
Jim Seay's Pole Tips and Swivel Fork	17
John Couty's 1834 Drawing of Bevil's Bridge	19
A Batteau in Action, with Two Steering Oars	20
A Batteau Built Too Big for the Locks	21
Map of Petersburg's Canal Basin	23
The Basin Mill	23
Diorama of Indian Town Creek Aqueduct	24
John Couty's 1834 Drawing of the Aqueduct	24
The Aqueduct after its Destruction in 1865	25
Another View of the Aqueduct in 1865	26
The David Bruce Stone	26
Ruins of the Aqueduct down Indian Town Creek	27
The Aqueduct Today	27
Map of the Aqueduct and Toll Locks	28
Jim Seay in the Toll Locks	29
The Four Toll Locks, from a Diorama	29

xii Illustrations

Close-up of Jim Seay's Ferry Boat	30
Diagram of Jim Seay's Ferry Boat Operations	32
Latrobe's View of the Appomattox in 1796	34
Jim Seay's Paddle in Olger's Store	37
Pocahontas Basin, Petersburg's Famous Pothole	39
Monument to a River Tragedy	41
Jim Seay's Dugout Canoe	42
Site of Frank Jay Gould's Mansion	44
Jim Seay and his Floating Cabin	47
Another View of Jim Seay's Cabin on Lake Chesdin	48
The Trolley Car at Ferndale Park	50
Ferndale Park's Spring-fed Fountain	52
Ferndale Park: The Store and Pool Hall	54
Ferndale Park: The Theatre	54
Ferndale Park: The Shooting Gallery	55
Ferndale Park: The Hobby Horses	55
Ferndale Park: The Swings and Bowling Alley	55
Ferndale Park: The Superintendent's Residence	56
Ferndale Park: The Restaurant	56
The Upper Appomattox Canal's "Abutment Dam"	57
Blasting-holes on the Canal	59
Aerial View of Brasfield Dam and Lake Chesdin	60
The Last Whitewater Trip to Brasfield Dam	62
The Mystery Hole in Lafayette's Rock	64
Jim Seay's Dam at Lake Margaret	68
"Built By J. W. Seay"	70
Olive Hill Plantation	72
A Lost Art Rediscovered at Olive Hill	74
Beadle's Dummy	76
Perspective View of Matoaca Mills about 1902	79
The Picker House at Matoaca Mills	79
The Office Building at Matoaca Mills	80
The 1834 Inscription at Matoaca Mills	80
Seward & Munt, North of Campbell's Bridge	81
"Not One Cent for Tribute"	82
Bellvue Plantation	83
Fold-out Map of Virginia's 19th-Century Navigations	99
Fold-out Map of the Upper Appomattox Navigation, and the Upper Appomattox Canal	101

PREFACE

I take great pleasure in expressing my personal satisfaction, and that of the Historic Petersburg Foundation, with the publication of this delightful volume. It represents a decades-long labor of love by Bill Trout and many others, but it has always been intended that the wry, expansive spirit of the subject of the book should not be smothered beneath the weighty scholarship and labor that have made the book possible.

Jim Seay's spirit, his love of nature's bounty along the Appomattox River, his curiosity and inventiveness, his joy in telling stories about himself and about the fascinating characters from his past and present who so joyfully peopled his life—this spirit comes through loud and clear in his own words, and in those of the many others who took such pleasure in knowing him. It is my loss never to have known him in person, but, with the publication of this slender volume, we will all feel his presence in our lives.

Bill Trout was one of the founders of both the Virginia Canals & Navigations Society (1977) and the American Canal Society (1972), and has been the President of the latter organization since 1983. Bill is surely the best possible person to put this book together, as his own wry sense of humor, and his own love of the river and of stories about odd and curious characters, perfectly compliment Jim Seay. Bill first talked to Seay in 1968, and then often explored the river with him until Seay's death in 1973. Beginning in the 1960s, Bill had been studying and writing about Virginia's canals and navigations, at first during vacations, then full-time starting with his retirement in 1982. One of the principal subjects of his study throughout this period was the Upper Appomattox Navigation. His first publication on the subject, "On the Future of the Appomattox Navigation," was printed in the *Progress-Index* on 11 June 1967. Recently, he has published *The Appomattox River Atlas* (1990), and he projects publication of *The Falls of the Appomattox Atlas* in 1993.

In 1988, Bill began transcribing the tapes made of interviews with Jim Seay by C. R. (Dick) Pitts in 1967 and by Mrs. Gretchen Woodruff in 1970. Beginning in 1989, he added interviews with others who knew and loved to talk about Seay. Bill and I have been working on the project together since he showed me his proposed book in 1989.

The Historic Petersburg Foundation recognized the importance of the project and voted in early 1990 to fund publication of *Appomattox River Seay Stories* through its Arthur W. Ritchings Memorial Publications Fund. It has been my great pleasure to work with Bill in our effort to bring the project to fruition. *Appomattox River Seay Stories* is the first-published book not only of the Ritchings Fund but of the Historic Petersburg Foundation. It is fitting that the book is being brought forth on the occasion of HPF's twenty-fifth anniversary. We trust that it will be the first of many publications. In fact, we have already projected several, including the new *Falls of the Appomattox Atlas* by Bill Trout, a book about Battersea, and a *Journal of Petersburg Architecture and History*.

Arthur W. Ritchings served as Secretary and then Executive Secretary of Historic Petersburg Foundation from 1971 to 1978, as its President from 1978 to 1980, and then again as President in 1984-85. A resolution passed by the Foundation board on 26 September 1988, on the occasion of Arthur's death, affirmed that Arthur "can never be replaced in our hearts"; that "his calm, kind, loving, and gentle ways brought us through many difficult and trying times"; and that "we have lost a part of our hearts and souls and are ever grateful that God made Arthur a preservationist and gave him to us as our best friend." Soon thereafter the Foundation established the Arthur W. Ritchings Memorial Publications Fund to honor his memory. I knew Arthur only in the last years of his life, but his impact upon those around him was such that I too will treasure his memory in my heart. May our publications be ever worthy of his name and spirit.

Arthur Ritchings, Bill Trout, Jim Seay—thus we begin our venture in publications designed to disseminate knowledge and appreciation of Petersburg's rich heritage. We hope that our first little book will bring to you, dear reader, as much pleasure as it has brought already to me.

R. Dulaney Ward, Jr.
4 September 1992

The Historic Petersburg Foundation

The Historic Petersburg Foundation, Inc., was founded in 1967 by a group of individuals, working with the encouragement and assistance of the Petersburg Chamber of Commerce, who were motivated by a desire to halt destruction of Petersburg's historic heritage, by a concomitant desire to utilize that heritage to provide the attractive power for heritage tourism, and by a perceived need for a local organization which could specialize in the acquisition and resale of historic buildings for adaptive re-use. The corporation was established with the express mission to "acquire, hold, improve, preserve, develop, and restore sites, buildings, residences, and squares which are part of the original plan of Petersburg, and to preserve the neighborhood design in mass and proportion, as well as other structures of historical or architectural interest in and around Petersburg; and to increase and diffuse knowledge and a greater appreciation of such sites and structures."

One of the major programs of the Foundation has been the purchase of endangered buildings and their resale to individuals who are interested in rehabilitating and restoring them. To insure that restoration occurs in a timely manner and that the buildings are preserved, protective covenants are placed on them. These covenants are recorded as part of the deed and remain effective in perpetuity.

Additional programs which the Foundation sponsors include the Battersea Ball, the Historic Homes Tour, the Preservation Awards, and the Preservation Conference, all conducted annually; the Easements Program, the Preservation Grants Program, Research and Publication, the Historic Petersburg Real Estate Fair, and a Series of Lectures, Workshops, and Exhibits.

Highlights of some of HPF's major accomplishments include saving from demolition and/or assisting in the restoration of Baltimore, Smith's and May's Rows, on High Street; the Carriage House Apartments and the Appomattox Iron Works, on Old Street; the Wallace House, the Jane McIlwaine House, and the Tatum House, on Market Street; Mayfield, on West Washington Street; the Caretaker's Cottage in Poplar Lawn Park; Strawberry Hill, on Hinton Street; Jones Mitchell's and Peter McCulloch's Rows on North Sycamore Street; Battersea; and the John Baird House and the Margaret Kenney House on Grove Avenue.

Historic Petersburg Foundation also assisted in the implementation of local historic zoning ordinances to protect our vanishing resources, and the creation of the Old Town, Centre Hill, Folly Castle, South Market Street, Court House, Poplar Lawn, and Battersea local historic districts.

Membership in HPF is open to all who are interested in our city and in the aims and objectives of the Foundation. HPF wants to know about anyone interested in what is being done to preserve the best in Petersburg for its future citizens. Please call HPF at (804) 732-2096, or write, 532 Grove Avenue, P.O. Box 691, Petersburg, VA 23804.

The Virginia Canals and Navigations Society

The Virginia Canals and Navigations Society, Inc., was formed in 1977 to preserve and enhance Virginia's rich inland waterways heritage in all its fascinating aspects. History, exploration, archaeology, modeling, local lore and legend, restoration, preservation, park and trail development—these are some of the many areas of interest our members pursue, to their own great satisfaction and frequently to the lasting benefit of their communities and state. The society is increasingly active as a focus for waterways interest, and as a collective voice of all those concerned about the future of this priceless part of our past.

The realization of Virginia's extensive canal and navigation system was brought about through the vision, genius, and faith of some of the most remarkable men of the age, including, not surprisingly, George Washington, who was president of both the Potomac and James River navigation companies. Unhappily, their creations have largely become the victims of neglect and public apathy, although here and there a canal park, trail, or scenic river has demonstrated the value of the old works as unique, eye-pleasing amenities and genuine historical memorials to our ancestors' ingenuity.

VC&NS members receive the illustrated quarterly *The Tiller* and other items of special interest; the opportunity to enjoy the company of others at society and local chapter meetings and field trips; and the chance to work together toward making our world a better place in which to live. Please join us! Write to the VC&NS Secretary, at 6826 Rosemont Drive, McLean, VA 22101, or to the permanent address, c/o the Alexandria Waterfront Museum, 44 Canal Center, Alexandria, VA 22314.

"I got all of my happiness in my life from what God created, the rivers, the hills, the woods, and those things."

Jim Seay at home at age 89. Photo by Tom Wessells in the December 24, 1967, **Richmond Times-Dispatch**. Courtesy of Richmond Newspapers Inc.

INTRODUCTION

"A Seay never would have settled in this part of the country, but the old grandfather tracked a rabbit down here from Louisville, Kentucky, in a deep snow, and the snow melted and the old man couldn't find his way back and he settled here."
(Blackford Smith, quoted by Jim Seay)

Jim Seay was a remarkable man. He was a time machine to a lost era of George Washington's day, when white-water rivers were the super-highways of commerce, when homes, towns and cities faced the river because that's where the action was. Along more than a thousand miles of Virginia's upland rivers, iron men in wooden batteaux poled their way, carrying down tobacco, flour, iron, and the products of the interior to the coastal cities, and brought back pianos, porcelain, and other articles of civilization from all over the world.

A century ago, in 1890, when he was only twelve years old, Jim Seay became a part of that lost era. For twelve years he was a batteauman, poling his boat down the Appomattox River and through its canals to the canal basin in Petersburg. He considered himself, in his own words, "the last man to run the boats."

We hereby present Jim Seay with the honorary title of "The Last of the Appomattox River Batteaumen," even though there are other claimants to that honor: the late T. D. Cox of Matoaca considered Willie Crowder and the Holt family to have been the last to work on the canal, after Jim Seay, and Seay didn't run his batteau all by himself. Hubert Holt, an expert on the Matoaca Mills where he grew up, recalls that Seay had no permanent partners in his batteau, ferry, and construction jobs, but worked with different men including Holt's cousin, his father Alpheus Holt, Tom Holt, Hugh Allen Andrews, and Hootie Bryant.

When the canal closed in 1902, Seay did not abandon his batteau but put it in the river at Matoaca. He built a cabin on it and used it as a ferry boat to carry passengers back and forth between Matoaca and Ferndale Park (now called Appomattox Riverside Park), until a bridge was built there.

Jim Seay was an outdoorsman all his life and could do most anything—house moving, blasting, concrete work—when he wasn't out on his river hunting or fishing. He had an enquiring mind and delighted in discovering and puzzling out the secrets of nature and of the past.

Jim Seay lived almost a century, from 1878 to 1973. He had a remarkable memory and was always ready to tell stories of his own experiences going back to his very early years, as well as stories the older batteaumen told him going back before that, and other nuggets of historical lore which he picked up.

Introduction

I consider it a great privilege to have known Jim Seay. He was my window to the batteau era in Virginia. He enjoyed telling me about the old days and took me out to the old canals along the Appomattox. Before he died, he gave me his relics from the batteau era—a batteau steering oar, two iron pole tips and a steering-oar swivel fork, for the Virginia Canal Museum which we hope to have someday. Until 1983, when canal buffs excavated the canal basin in Richmond and found sunken batteaux and more pole tips and forks, these were the only batteau artifacts known in Virginia. Thanks to Jim Seay, we knew what to look for in the canal basin mud. The 16-foot-long steering oar, now on display in Reynolds Metals Tidewater Connection Locks Park at 12th and Byrd streets in Richmond, is still the only one we have from the Batteau Era.

Virginia is now enjoying a new Batteau Era, this time for recreation. After we excavated the batteaux in 1983, Joe Ayers built a replica, navigated it down the James, and started the annual James River Batteau Festival. Now there are two dozen operating batteaux in Virginia, perhaps more than in Jim Seay's time, and several hundred intrepid men, women, and children have tried to rediscover the lost art of batteau poling and navigation. Now that we have batteaux to play with, we have many more questions to ask of Jim Seay, but it's too late.

Fortunately for posterity, several hours of interviews were tape recorded by C. R. (Dick) Pitts, Jr., in 1967, and by Mrs. Gretchen Woodruff in 1970. They have given me copies of the tapes and permission to use the best of the stories in this collection. Without their forethought this little book could not have been written. All the stories in this book are from these tapes of Jim Seay, unless otherwise noted. I've also recorded a few stories about Jim Seay from Ferndale resident Hubert Holt; from Jim Seay's son Frank, who still lives in Matoaca and has inherited his Dad's gift of curiosity; and from Philip R. Roper, Jr., who first knew Jim Seay 67 years ago when he built Lake Margaret Dam. We are also indebted to Gary Dalton, who wrote an excellent article about Jim Seay in the *Richmond Times-Dispatch* which is reprinted here.

Over the years, many others also contributed encouragement, advice, and their expertise, including my mother and (alphabetically) Dr. James H. Bailey, Fred R. Bell, Nancy Dunnavant, William J. Graham, Catherine H. Grosfils of Colonial Williamsburg, Richard D. Hartman, Theodore W. Haxall, Dr. Joseph C. Hillier, Richard L. Jones, Bill Martin, Carlton N. McKenney, Jeff O'Dell, Warren C. Purcell, Russell Rulau, James M. Seay (Jim Seay's great-grandson), Mr. and Mrs. Rubin Traylor, and Dulaney Ward.

For more about the history of the Appomattox and its canals and locks, see **The Appomattox River Atlas**, by W. E. Trout, III, published in 1990 by the Petersburg Department of Tourism for the Virginia Canals & Navigations Society.

We'd like to hear from anyone interested in river history, for there's plenty of work for us all. We'd also like to know about other tapes, photographs, articles, and artifacts about Jim Seay, the Appomattox, or the batteau and canal eras in Virginia. Please write to W. E. Trout, III, 35 Towana Road, Richmond, VA 23226, or to the Virginia Canals & Navigations Society, c/o the Alexandria Waterfront Museum, 44 Canal Center, Alexandria, VA 22314.

Bill Trout, March 1991

RIVER IS FOCAL POINT OF A HAPPY LIFE:
A BIOGRAPHY OF JAMES WASHINGTON SEAY, 1878-1973

by Gary Dalton
*Edited from the **Richmond Times-Dispatch**,*
December 24, 1967, Section F, pp. 3-4

"I've had a happy life, I got all of my happiness . . . from what God created—the rivers, the hills, the woods, and those things . . . . I never cared about dancing and music and frolicking and that kind of stuff. But right in the river, the wildest woods you ever saw, was where I liked best."

Lake Chesdin, the 17-mile-long lake impounded by the new Appomattox River Dam near Matoaca, covers a lot of memories for James Washington Seay.

For 12 years, summer and winter, Seay operated a batteau on the Appomattox River. The time was 1890 to 1902 and "Jim" was only 12 or 13 when he started.

Today [in 1967], Seay is 89 and gray-haired and he doesn't get around in the woods as easily as he once did. But he remains strong and erect and his memory would put to shame many men half his age.

In between his youthful days on the batteau and his present semi-retirement, Seay has operated a ferry, worked in a foundry, repaired bicycles, built dams, dug wells, and moved houses.

Almost everyone in the Matoaca community knows "Mr. Seay." He and his parents were here when the community consisted of two or three houses and you either walked or rode horseback the five miles into Petersburg.

To some he's known as the "river man," testimony to his years of working, fishing, and hunting on the Appomattox, which flows within a quarter-mile of his home.

"Planned to cut me some firewood."

Jim Seay at home beside his 1904 gasoline-powered saw. (Photo by Fred Van Deventer in the November 21, 1971, Petersburg **Progress-Index**, courtesy of the Chesterfield Historical Society)

That river has been one of the focal points of his life, from the time he was able to fish from the riverbank, sip water from its springs and watch the batteaux pass by.

Seay is, he says, the last survivor of the men who poled those boats up the Appomattox River.

He recalled recently that he began working on a batteau in 1890 when he was 12 years old. For the first few trips, he rented a boat, but then he bought his own for about $150.

For most of the next 12 years he hauled cordwood, lumber, railroad ties, and pulpwood 30 miles down the river and through the Toll Locks of the Upper Appomattox Company to Petersburg.

"You could go down them (the Toll Locks) free," he said, but "you had to buy a ticket in town to come back and they sold for $10 a dozen. I only bought a half-dozen at a time."

The canal company had its beginning in 1795 when it was authorized by the General Assembly to build the Upper Appomattox Navigation, a series of canals and other navigation improvements, between Farmville and Petersburg, a distance of 100 miles. It eventually extended to Planterstown, which no longer exists, 23 miles further upstream of Farmville.

Although records of boat travel on the upper Appomattox River are almost non-existent, Seay thinks that boats were going all the way to Farmville before the building of the canals. However, because of the varying water levels of the river, water travel was unpredictable, so the Upper Appomattox Company was authorized to build a series of dams with canals and locks to raise the water level at shallow spots and to bypass rapids.

But in 1854 the Southside Railroad (later part of the Norfolk and Western) completed its line between Petersburg and Lynchburg, via Farmville, and killed most traffic on the river between the two towns.

The canal company persisted, however, and continued to serve farmers along the river for about 30 miles upstream, near Bevil's Bridge (Rt. 602) and the mouth of Deep Creek.

Unlike the Kanawha Canal along the James River, and the Chesapeake and Ohio Canal along the Potomac, the Upper Appomattox Navigation primarily utilized the river itself as the navigation channel. The canals along the Appomattox, according to Seay, were seldom more than a mile long. When the batteaux reached a canal's end, they would go into the river for two or three miles before returning to another stretch of canal.

Boats on the Appomattox River were poled by two men instead of being pulled by mules as on the Kanawha and C&O canals, which had continuous towpaths.

When the boats went upstream, a man on each side of the boat would stick his pole in the river bottom and walk from the front to the rear of the boat, moving the boat forward as he walked. A third man had to stand in the back and steer with a long wooden steering oar.

The only canal with a towpath on the Upper Appomattox Navigation was the Upper Appomattox Canal, which ran for five miles from a dam to a canal basin in Petersburg, passing down a series of four locks known as the "Toll Locks." The crew usually pulled the batteau along the canal, but, on one occasion, he said a man wanted him to haul a load of sheet iron up the canal and

"These boats I'm telling you about hauled merchandise and produce from the farmer on the riverbank."

No examples of batteaux existed in modern times on the James and Appomattox until 1983, when dozens of sunken batteaux were discovered in Richmond's canal basin during the construction of the James Center. One end of the sixth boat to be dug out of the mud is shown above. Archeologist Lyle Browning is walking along one of the two planks from which the boatmen poled the batteau nearly two centuries ago. (Trout photo)

Seay hitched a mule to the boat. The mule wouldn't pull at first, Seay said, but after two poles got the boat moving, the mule picked up the slack.

Seay said that each of the canal locks was eleven feet wide and 74 feet long, with gates that closed in a "V" shape pointing upstream. Each lock would raise or lower a boat about 8 feet.

The boats were six feet wide and 60 feet long.

"When you ran a boat in [a lock], you hardly had room enough to turn the water in without water pouring down in the boat," he recalled.

Although many of the old locks have been covered up by Lake Chesdin, which finished filling only last week, the remains of perhaps the most interesting still exist in western Petersburg.

The Toll Locks form a four-lock staircase, one of the three stone locks on the river (the rest were wooden) and had four lock chambers, carrying the canal down a hill like four stairsteps, lowering boats a total of 33 feet. Parts of the two lowest chambers are still visible aboveground.

"The most dangerous thing was for the man operating the [Toll Locks] to know exactly how much water to let in. If they didn't have enough water coming in from the upper gate to accommodate a leakage in the lower gate, when your boat was leaving one chamber and going into another, it would hang up." Just down the canal from the Toll Locks are the remains of Indian Town Creek Aqueduct, a large stone archway which carried the canal over the creek.

According to Seay, a flood in 1865 carried away the arch of the aqueduct and it was rebuilt of wood. Every time his boat crossed the wooden aqueduct, Seay said, it would hit against the wooden sides and cause the structure to shake.

Today, only two massive stone walls, about six feet wide and 20 feet high, remain to show where the aqueduct had been.

The canal closed about 1902, Seay remembers.

"Then I took my river boat that I was operating and put it down here [at Matoaca] across the river as a ferry and ran that until they built a bridge [now Rt. 600] and I had to discontinue my ferry business."

After that, Seay had a variety of jobs, including working for the Petersburg Iron Works and a bicycle shop. He recalls that he quit the iron works job to work in the bicycle shop because he could get a bicycle and wouldn't have to walk the five miles between Matoaca and Petersburg as he had been doing.

In addition, he has trapped, dug wells, moved houses, run a saw mill, and dragged bodies out of the river. However, he calls cement work his trade and has built dams, pools, and basements. He still installs sidewalks and other minor projects and does not consider himself retired.

But he takes it easy, enjoying his daily milk toddy for breakfast, fishing, and four or five cigars a day.

"I've had a happy life," Seay said. "I got all of my happiness in my life from what God created, the rivers, the hills, the woods, and those things.

"My greatest happiness in my life was through them. I never cared much about dancing and music and frolicking and that kind of stuff. But right in the river, the wildest woods you ever saw, was where I liked best.

"When I was a young boy I used to cater to this river up here, but after I grew up, me and my sons would go to the Nottoway River to hunt and, oh my God, that was the huntingest ground on this earth. We would go moonlight

6   *Appomattox River Seay Stories*

"He took it down, plank by plank, and transported it by horse and wagon to his homesite in Matoaca and rebuilt it."

Jim Seay's home in Matoaca, previously the home of Petersburg merchant Anthony Rosenstock, stood in its earlier incarnation on the east side of Market Street, just south of Washington Street, in Petersburg. Since this photograph was taken the house has once again been demolished. (1977 photo courtesy of Jeff O'Dell, from his book, **Chesterfield County: Early Architecture and Historic Sites***)*

nights and drift down the river and shoot those coons; oh my, it was something in this world."

Seay doesn't hunt any more, although his eyesight is good enough to read a newspaper without glasses and write a neat hand. He lives quietly in his two-story, weather-beaten, frame house on River Road.

His few concessions to the modern way of life are electric lights, an electric refrigerator, a radio, and his pickup truck.

His water comes from a hand pump in a well Seay dug in 1912. A wood-burning range furnishes heat for the kitchen and the little cooking he does.

His house at 1011 River Road is probably as old as he is. Seay bought it in Petersburg in about 1912 for $300. He took it down, plank by plank, and transported it by horse and wagon to his homesite in Matoaca and rebuilt it.

"It was one of the finest buildings there was in Petersburg," Seay said. "It was known as the Rosenstock home" for the same family that has an interest in Rucker-Rosenstock department store, one of Petersburg's largest.

Scattered about the house are mementoes of Seay's life. His hobby is collecting Indian relics and he has dozens of arrowheads and about a dozen stone axes found on his walks and hunting trips.

Under the house is the 12-foot steering oar he used on his batteau.

In a shed behind the house are the remains of an Indian dugout canoe that he retrieved from the mud of the Appomattox River. Seay will show visitors two bullet holes in the canoe, one where the bullet went in and another where it came out.

He has given other canoes to his sons and sold his best one to Chesterfield County for display in its museum.

Another memento is the end of a pole that was used on a batteau more than 100 years ago. Seay found the pole stuck in a crack in a rock in the river about 10 years ago.

That portion of the river was bypassed by one of the canal company's canals in the 1830's, so hadn't been used by batteaux for more than 100 years, Seay said.

Recently, Seay completed work on a 12-foot boat which he built to row across Lake Chesdin. It is one of many boats he has built since his boyhood days.

Back around 1915, Seay said, he and a friend built a houseboat, shipped it to Farmville, and drifted back down the river to Petersburg. At one spot in the river they shot 11 mink, which, at $12 each, more than paid the $15 or $20 it cost to ship the boat to Farmville.

Later, in the '30s, Seay built a little cabin on the river and "called it my houseboat. It was an old boat and after the boat sprung a leak, I pulled it on the highlands and made me a cabin."

In 1940 a flood washed away his cabin, but he built another one that year on an island in the Appomattox River. It was made of cut timber and was covered with sheet metal.

That cabin has lasted until today, despite the new dam which flooded the island where it was located.

Seay said, "When they built this lake, I asked the gentlemen at the water authority if I could preserve my house by floating it away on barrels and they gave me the privilege. So I put 24 barrels under it, and now it is floating on the water like a cork.

"That part of the river hadn't been used since 18 and 34."

Low Wall Sluice, by-passed by Caudle's Canal in the 1830's, is the only batteau sluice with a hauling wall still visible on the Appomattox. Going upstream, batteaumen got out onto the hauling wall and towed boats up through the fast water; it was easier than poling. This is a view upstream at low water, with the wall to the left and Lake Chesdin Dam in the background. Canoes still use this sluice, which is just downriver of Chesterfield County's Appomattox River Canoe Launch. (Trout photo, 1990)

"Now, possibly, I will carry it to the extreme upper end of the lake and place it somewhere on the old original river again so that I can see it. I would rather be on the old river above the lake because it looks more like what it did look like.

"I like to remember it as it was."

* * * * * * * * *

Jim Seay, born in Matoaca on June 27, 1878, passed away there on March 19, 1973, at age 94, and was buried in East Matoaca Cemetery. Surviving him at that time were three daughters, Mrs. Audrey Vaughn, Matoaca; Mrs. Rosa Meadows, Ft. Ashby, West Virginia; and Mrs. Virginia Meadows, Matoaca; three sons, James M. Seay Sr., Matoaca; Gordon E. and Franklin R. Seay, both of Petersburg; 13 grandchildren, 23 great-grandchildren, and three great-great-grandchildren. (*Progress-Index*, March 21, 1973)

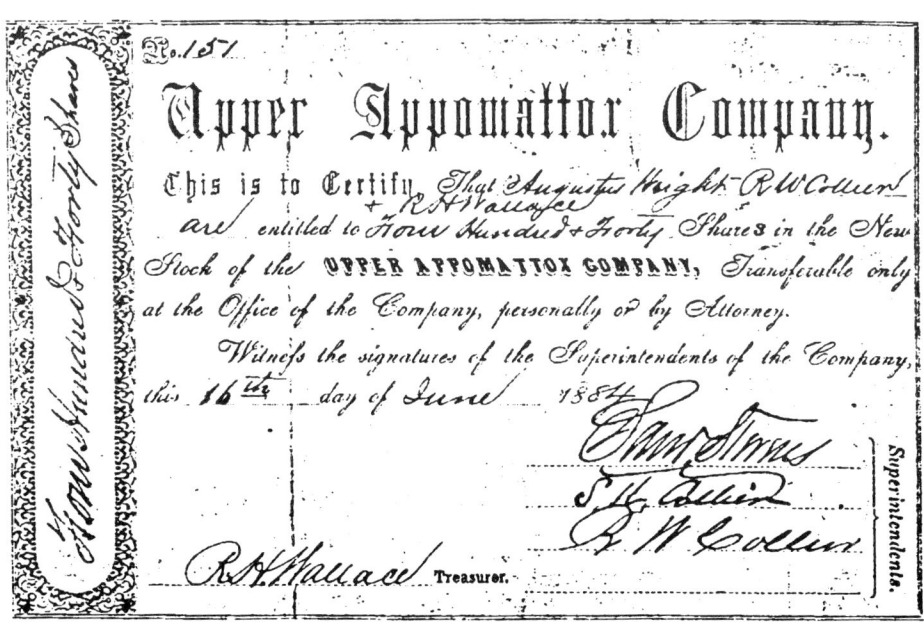

A rare example of an Upper Appomattox Company stock certificate, dated 1884. (Richmond Public Library clipping file)

After batteaux were rediscovered in Richmond's Great Basin in 1983, over two dozen replicas were built by volunteer groups. One of the first was the **Lord Chesterfield**, *which is affiliated with the Chesterfield Historical Society.*

Re-creating a scene from the original Batteau Era, the **Lord Chesterfield** *lies moored in the Upper Appomattox Canal in Ferndale Park on Batteau Day, October 15, 1988. The* **Lord Chesterfield** *was the first batteau in modern times to navigate the Appomattox. Note the long steering oars at each end and the tobacco hogshead.*

(From a water-color painting by Art Markel, in **The Appomattox River Atlas**, *1990, published by the Petersburg Department of Tourism for the Virginia Canals and Navigations Society.)*

A Biography of James Washington Seay 11

"I read a book one time of some Frenchman, a celebrated Frenchman came over here and made that trip down the river."

*In June, 1796, architect and engineer Benjamin Henry Latrobe explored the Appomattox River for the newly-created Upper Appomattox Company. This pioneering voyage was closely duplicated two centuries later by the batteau **Lord Chesterfield**, shown during her launching in Farmville on September 25, 1987. The river was barely wide enough for the 50-foot long boat. (Trout photo)*

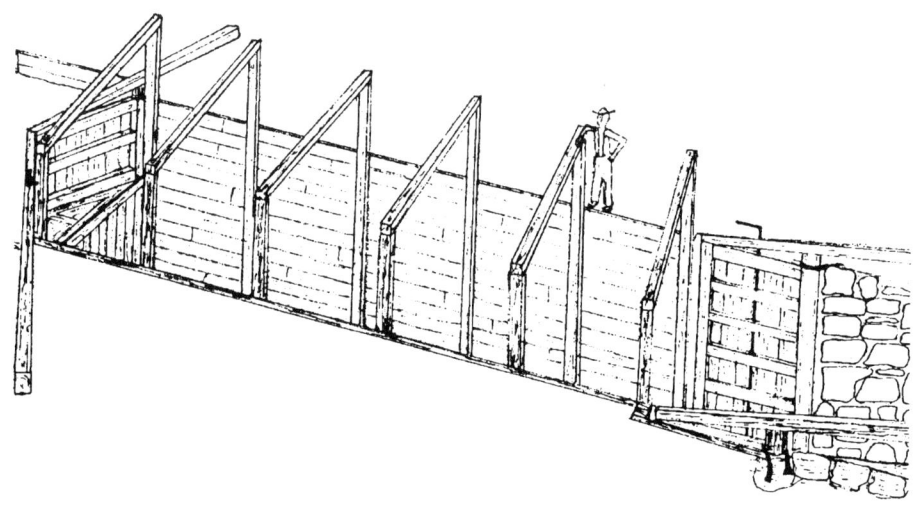

"I remember when a lot of the locks were still standing . . . 200 feet above [Caudle's Lock] is where they're building the big dam."

Above:
A drawing of Caudle's Lock, based on Jim Seay's description. The wooden lock walls were braced apart by overhead cross-timbers. This site can still be found downriver from Lake Chesdin Dam but the walls decayed aboveground long ago, leaving a depression in the ground. (Drawing by Trout)

Opposite, top:
Jim Seay in 1968, beside the scale model he made of Caudle's Lock. Appropriately, the background is the stone wall of the Picker Building at Matoaca Mills, built at the same time as the lock—1834.

Opposite, bottom:
Another view of Seay's lock model, held by Mrs. W. E. Trout, Jr. (Trout photo)

A Biography of James Washington Seay 13

"Going upstream you throw that pole out into the water and . . . walk the length of the boat."

"Farmville at Appomattox River," by Lewis Miller (1796-1882), the only known painting of an original Appomattox River batteau. On a trip in 1853 the artist left the train (probably at High Bridge) to experience a four-mile batteau trip up to Farmville, where he met the stage. Note the men poling. Did the artist use his imagination when he drew a rudder, instead of a steering oar, at the stern of the boat? (Courtesy Abby Aldrich Rockefeller Folk Art Center, Colonial Williamsburg Foundation)

CHAPTER 1: JIM SEAY, BATTEAUMAN

"I was telling you about I was the last man to run the boats."

HOW JIM SEAY POLED HIS BATTEAU

"The boats was operated by a pole, and an iron spike on the end of the pole. And the boats was 60 feet long and a walk board all the way along on the inside of the gunwhale of the boat. . . . Didn't use no poles coming downstream, going upstream you'd throw that pole out in the water and put it against a pad at your shoulder, then you would shove right on down with that pad, walking right down the line until you walked the length of the boat. Had a walkway all the way inside the boat. Then when you turned and went back the boat would have the same momentum, hadn't lost no momentum at all when you walked back to get another holt.

"And when I operated the boat I hauled wood and logs and things of that kind but these boats I'm telling you about [in the old days] hauled merchandise and produce from the farmer on the riverbank. Every farmer had a wharf or landing to send his stuff into town. And one of the old men when I was a boy was watching me clean my boat out when I was hauling some cordwood down there. And he said, 'You got cedar poles in your boat.' I said, 'Yes, Sir.' He said, 'Well, we never did have cedar poles, we always used pine poles. Cedar poles got crooked. We used pine poles.'"

HE DISCOVERS
AN EARLY BATTEAU POLE

"Not too long ago I happened to be around on the back side of the river where these canals were dug in '34 and I saw a splinter standing in this place with swift water, wiggling like that. I said, 'Wonder what holds that splinter there?' And it was a fall, or sluiceway where the old boats used to run, and evidently a man with his pole got this thing, his pole, hung in a crack in the rock and broke his pole. And it stood there for a hundred years. . . . Pine, that was what the old first men used. I used cedar but it was always crooked. But he used pine, never was crooked. And I went to look at the thing close and it was tight. And I pulled it out and the iron spike from the pole was in a crack in the rock. . . . Isn't that something? The spike made by a blacksmith. . . . And that part of the river hadn't been used since 18 and 34 . . . because no boats had travelled in that section of the river since the canal was built."

"A man got his pole hung in a crack in the rock and broke his pole."

This woodcut of an exciting moment on the New River, from the February 21, 1874 issue of **Harper's Weekly**, *dramatically illustrates the use of a batteau steering oar with a swivel fork, and iron-tipped poles for fending off rocks. Going upstream, the poles were used to propel the boat.*

Jim Seay, Batteauman

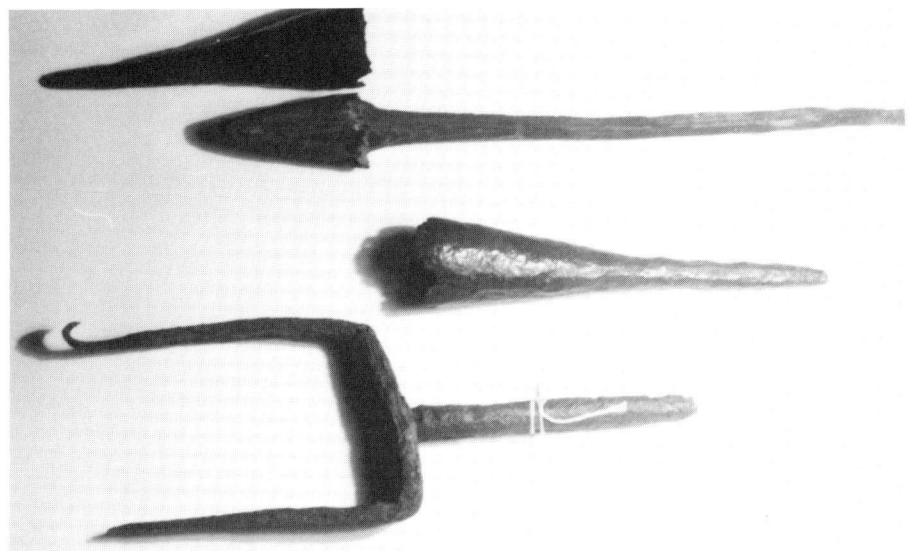

"It stood there for a hundred years . . . and I pulled it out and the iron spike from the pole was in a crack in the rock. . . . Isn't that something?"

These two metal-tipped ends of poles used by batteaux on the Appomattox were found by Jim Seay. The one at the top was found with the end of the wooden pole still attached (separated in photo), caught between two stones in a rocky place in the river near the upper end of Caudle's Canal. The boatman must have caught his pole between the rocks, and it broke off.

The other pole tip was found lower down at Broad Rock Falls. Because this part of the river was bypassed by Caudle's Canal in the 1830's, it probably dates back before then.

The fork was the swivel for a batteau's long steering oar. This one was found by Jim Seay near his cabin at Black's Wall. One of the prongs was bent; Seay thought the batteauman had dipped or swung his steering oar too far while negotiating the four-foot drop there, resulting in a catastrophe and the loss of his fork.

These rare artifacts were donated by Jim Seay to the Virginia Canals and Navigations Society, and were stabilized by the Virginia Department of Historic Resources. (Trout photo)

HOW AUNT HARRIET'S FURNITURE CAME TO MATOACA

"That bridge that crossed the transportation canal over yonder had to be high enough for a loaded boat to go under there and probably the load was around six foot above the level of the water in the canal. And all of the bridges that cross this whole river had to be high enough for boats to go under which was loaded, coming from Farmville and going to Farmville.

"And in 18 and 34 when the [Matoaca] mill was built, my old Aunt Harriet, who was minding me when I was a small boy, up until I was three or four years old, used to tell me all these things, and she said that they came from Prince Edward County near Farmville, and they came here to work in the mill. They were advertising for people all over the country to come to work in the cotton factory. And sure enough, they did, they come from all over the country. And these five brothers, my father and his four brothers, all came here about that time.

"And Aunt Harriet said that when they got to the river at a place known as Clement Town, which I have a picture of the mill now, that I took in 1907. And that's where Aunt Harriet and her father and her mother got on a riverboat [a batteau] to come here, to live here at Matoaca and work in the mill. And she was a young girl. And she said when the boatman came down the river and her father asked about taking his stuff on to Petersburg, he said, 'No, I can't because I'm already loaded.'

"And the river was high at the time and the bridges was just probably high enough at extreme low water to accommodate the boats going under which was loaded, so she said he agreed to put their furniture, what little bit they had, on top of the already loaded boat, and said every bridge you got to, had to take the furniture off, put it on the bridge, take the boat under the bridge, and reload it! And they had to go to Petersburg and come back up the road [to Matoaca] because [then] there was no way to cross the river, to get the furniture back out here!"

MR. BOWMAN
AND THE BARREL OF WHISKEY

"That's as far as the boats went, to that basin [in Petersburg]. Then they unloaded and took on another load for Farmville. One old man was a boatman on one of the boats. He didn't have any education. And when the agent in Petersburg gave him a bill of lading of what he had on board to take to Farmville, there were some barrels of whiskey on there. His name was Bowman—kin to these Bowmans you know right around here—Oscar Bowman was his name. So he put the bill of lading in the boat and he rode off with it and the boat load of stuff for Farmville.

"When he gets to Farmville, he didn't know that a gauge could tell whether he had robbed that barrel of whiskey and put water in it or not. When the man put his gauge in there he found that the whiskey didn't have the same gauge in it as it did when he left Petersburg. So he says to Mr. Bowman, he says, 'Mr. Bowman, I find that you got five gallons of old Appomattox River water in one of these barrels.' He said, 'What did you say the name of the river was?' He didn't deny taking the whiskey!"

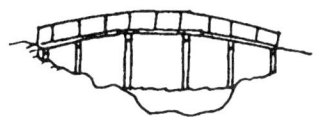

"And all of the bridges that cross this whole river had to be high enough for boats to go under."

Bevil's Bridge on Route 602 looked like this in 1834 when Aunt Harriet's furniture came to Matoaca. Drawing from John Couty's 1834 Survey Book in the Virginia State Library and Archives.

THE BOATMAN'S LOST GOLD

(The following two stories are not from tapes, but from notes of interviews with Jim Seay in the 1970s)

Some time before the Civil War, a batteauman carried a boatload of tobacco from Clement Town down to Petersburg. With a purse full of gold coins from the sale of his cargo, he began poling back upstream. On the way, the river began to rise; by the time he had locked into Black's Canal, ten miles above the basin, the river had become too swollen to work against, and he was feeling right poorly anyhow, so he laid over in the canal. Beginning to worry about his health on that stormy night, he hid his purse under a rock near his boat, and started walking toward the nearest habitation. But Alas! He died, never to return, and ever since, people have been looking for his treasure. Now here is where a knowledge of the canal helps, for the batteauman would have tied up in one of the laybys, and Jim Seay knew there was only one in that part of Black's Canal. Near there, under a rock somewhere, is the boatman's treasure. And—by the way—this part of the river is now twenty feet or so under the surface of Lake Chesdin, a fitting end for such a story!

"Didn't use no poles going downstream"

This batteau, descending the New River near McCoy, Virginia, has a sweep (steering oar) at each end. Detail from "Great Falls, New River," by W. L. Sheppard, from **Picturesque America**, 1872.

THE BATTEAU BUILT TOO BIG FOR THE LOCKS

It seems that in the late 1880s or '90s, Tom Hardy and David Wyatt built, in a boatyard near the head of the Upper Appomattox Canal, a batteau with unusual specifications for a man named Bosher, from Newport News. This boat was built on the plan of the traditional batteau, with one exception: Mr. Bosher wanted to carry a bigger payload every trip, so the sides were made higher than usual. Well, a man named Marick operated the boat for Bosher. He rather quickly found, to his dismay, that when the boat was carrying the dreamed-of bigger load, it settled so far down in the water that the canals were too shallow for it, and even worse, the boat couldn't scrape over the upstream end of the locks—over the miter or "mud" sill at the entrance.

The batteau was used for a long time after that, but always with a normal load like everybody else. It was still in perfect condition in the 1930's, when, tied up at Black's Wall (a relic of the early sluice navigation), it was allowed to sink and should be there still. Lake Chesdin has since flooded the site, but we hope someday to study this boat, the only remaining Appomattox batteau we now know of.

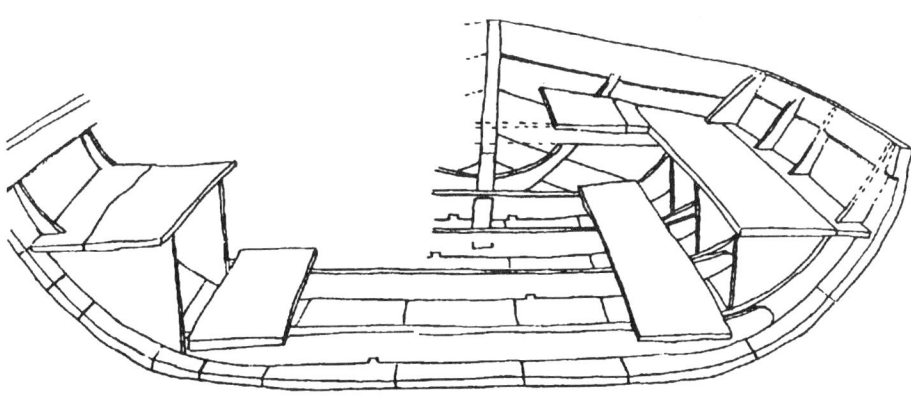

"The batteau built too big for the locks"

Mr. Bosher's batteau has not yet been rediscovered, but it could have been something like this one from the Rappahannock Navigation, recovered by Randy Carter and now in Warrenton's museum. Usually, batteaux had only two walkboards, one along each side (flat on the floor) which the boatmen walked on while poling. This Rappahannock batteau had sides so high that a second pair of walkboards, like benches, were added higher up to walk on. (From **The Rappahannock Scenic River Atlas***, 1992, available from the Virginia Canals and Navigations Society)*

THE CANAL BASIN IN PETERSBURG
AND THE MILLS THERE

"At the terminal, down at the end at South Street, they had a basin that was capable of turning many boats around and parking boats in, and a big shed where the produce brought from Farmville was placed in this shed. And the shed, the big house, had a long shed on it that boats would get under the shed to unload. And then right at the end of that where the water started from there to the river, sixty-foot fall from there to the river.

"I was going to tell you now about that old basin. Right at the edge of the old basin, the very first opening where the water left the basin towards the Appomattox River to go in the River at Campbell's Bridge, the silk man John N. Sterns first operated what was known as the Basin Mill, a great big wheat- and corn-grinding mill that operated on this water power that was going down that sixty-foot to the river.

"Well, then the water operated this building which was a tremendous big thing, I think it was three floors high, and ground wheat and corn. You know how High Street goes around there and becomes West High Street. Right on the Seaboard Railroad—that was the canal. That low place on West High Street where you go under the railroad bridge, that was known as the Basin Mill.

"Well, on that side they had water fall enough to create another plant, and that big plant was the same kind of corn or wheat mill was known as the Roller Mill. Right down Canal Hill where that shirt factory is now—or whatever it is—was the Blue Ridge Cotton Factory. [The canal water] operated those three plants, right there. And then it went on down under the Norfolk and Western Railroad and contacted a canal that the Munt Mill and the Poole Mill built together. And they had a canal on either side of the river flowing down to operate Munt's Mill and John W. Poole's Mill. And later John N. Sterns bought that property up and built a silk mill there. It also aided in running that. That's how much that water did after it had formed that transportation line all the way from Farmville down to [the basin in Petersburg]."

"When you got down to Petersburg you were at the head of High Street . . . at the old basin. That's where you turned the boat around at, right there."

Opposite, above: A map of the square containing the canal basin, showing the Basin Mill. The navigation canal came in from the left, and the overflow powered the Basin Mill and a series of others on its way down the hill to the river. The basin deserves protection as a unique time capsule of Petersburg's industrial past.

Opposite, below: A photograph of the Basin Mill, looking west across the basin itself, beyond the mill to the left. No boat or water is visible. (Map and photo from the Virginia Consolidated Milling Company scrapbook compiled by Charles Hall Davis, Dec., 1905, courtesy of the Petersburg Public Library)

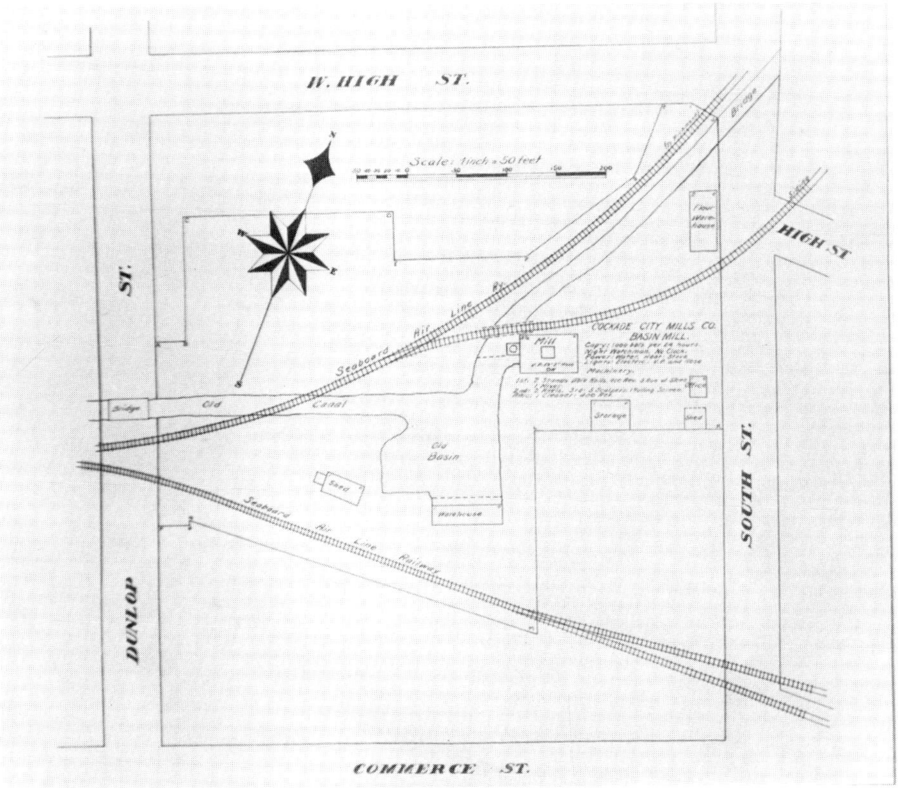
W. HIGH ST.
Scale: 1inch = 50 feet
Flour Ware-house
HIGH ST
COCKADE CITY MILLS CO.
BASIN MILL.
Mill
Seaboard Air Line Ry.
Bridge
Old Canal
Old Basin
Shed
Warehouse
Storage
Office
Seaboard Air Line Railway
SOUTH ST.
DUNLOP ST.
COMMERCE ST.

BASIN
ROLLER CORN MILL.

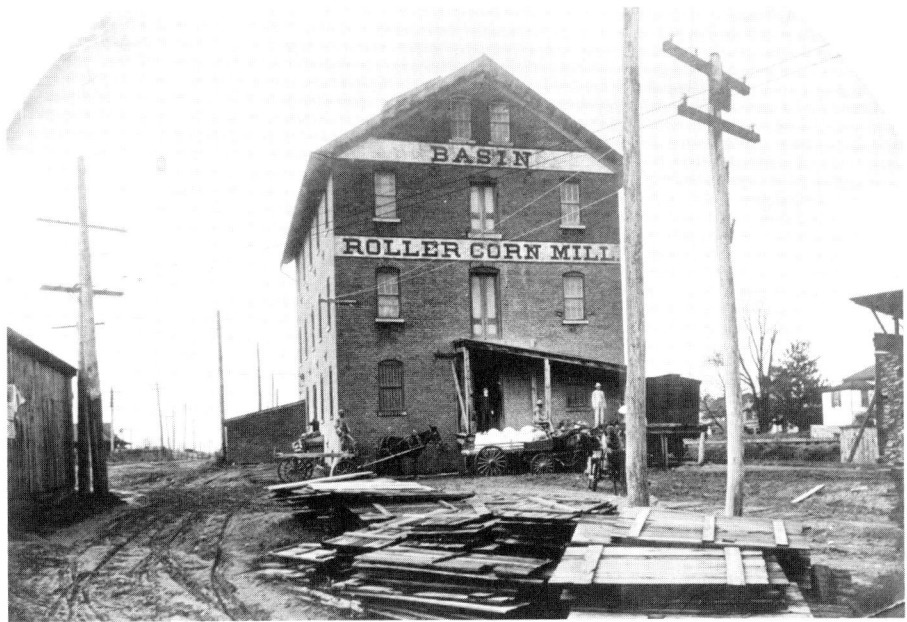

HOW THE CANAL AQUEDUCT FELL DOWN

"They didn't have men enough to form the lines back here [during the siege of Petersburg] to prevent the Yankees from taking Petersburg. So they built a big dam up there [on Indian Town (or Rohoic) Creek, to make a water barrier half a mile long], . . . named it Rohoic Dam. And it broke. And it washed away everything between there and the river. It washed away the canal, it washed away the Norfolk and Western [then the Southside] Railroad.

"Then they rebuilt the canal with a wooden flume, built out of wood—timbers—which before that was built a great arch out of rock cobblestones—one of the biggest jobs I ever looked at in my life to see where men could do without anything at all, no power at all. They built that arch over that Rohoic Creek out of cobblestones and that thing stood there until that dam broke and washed it away. Then they put that trestle out of wood cross there and I carried my riverboat through there. And if it bumped the side that would shake the whole thing. You were sure that it would break down!"

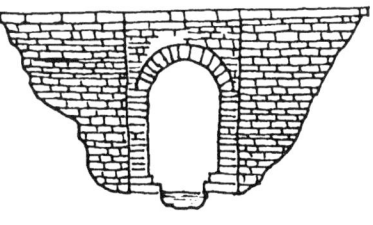

"They built that arch over Rohoic Creek out of cobblestones and that thing stood there until that dam broke and washed it away." Indian Town Creek Aqueduct (built of cut stone, not really cobblestones) as it was before the war, from a diorama (left) based on John Couty's 1834 drawing (right) and his survey. The stone arch carried the canal across Indian Town (Rohoic) Creek.

"They put that trestle out of wood 'cross there and I carried my riverboat through there. And if you bumped the side, that would shake the whole thing. You were sure that it would break down."

This and the following two photographs are the only early images known of Indian Town Creek Aqueduct, taken soon after the stone arch had become a casualty of the Civil War. A mile up the creek, the Confederates had built Lee's (or Rohoic) Dam, over a period of 15 days in August and September 1864. The pond was one of several, called "water men," which helped augment the defenses of Petersburg. In February, 1865, after a period of heavy rain, Lee's Dam burst and the resulting flood destroyed the stone arch of the aqueduct. Remains of the massive earth dam can still be seen on Rohoic Creek just north of the Boydton Plank Road (Rt. 142).

This photograph, taken in April 1865, probably by A. J. Russell, shows the massive scale of the stone abutments of the aqueduct—see the man perched on the stone ledge. The building in the background may have been the lock-tender's house. (Library of Congress Photo)

26 *Appomattox River Seay Stories*

This photograph, probably taken by A. J. Russell about the same time as the other photos, also shows the aqueduct's massive stone abutments, which are still standing today but desperately need stabilization. The canal must have been considered useful during the war, because by the time the photograph was taken in April the stone arch had been replaced by a wooden trough to carry boats across. In the 1890's, when Jim Seay took his batteau across, it was ready to fall down. Beyond the trough, looking down Indian Town Creek toward the north, is the Southside Railroad Trestle. (Library of Congress photo)

Opposite page, top: This view, taken from the ruins of the aqueduct in May, 1865, shows the stones of the washed-out arch strewn about downstream. Further down Rohoic Creek is the Southside Railroad's (now Norfolk Southern's) wooden trestle, since replaced by a concrete arch. (Quartermaster Corps photo, National Archives)

Opposite page, bottom: Part of the massive ruins still standing today on the west bank of Rohoic Creek . (T. T. Brady photo, 1988)

Below: One of the stones from the aqueduct was inscribed "DAVID BRUCE, 182__," presumably by the engineer or contractor who rebuilt the aqueduct of stone, after it washed out the first time, in 1826. (Stone rubbing by W. E. Trout)

Trestle work on Southside Rail Road
[illegible] ruins of [Rice?] Dam, May 1865

"You could go down [the Toll Locks] free, but you had to buy a ticket in town to come back."

Opposite page, top: *Jim Seay climbing up into one of the four Toll Locks in 1968. The left-hand wall has since collapsed. (Trout photo)*

Opposite page, bottom: *The four toll locks as they may have been in canal days, from a diorama based on John Couty's 1834 survey.*

"One of the biggest jobs I ever looked at in my life."

Below: *The massive ruins of Indian Town Creek Aqueduct and the Toll Locks which impressed Jim Seay are still there, preserved by Virginia Power. This map of the site was made in 1834 as part of John Couty's river survey, and is taken from page 120 of his field book in the Virginia State Library. Added to the map in dashed lines are the present-day road, railroad, and VEPCO dam across the canal.*

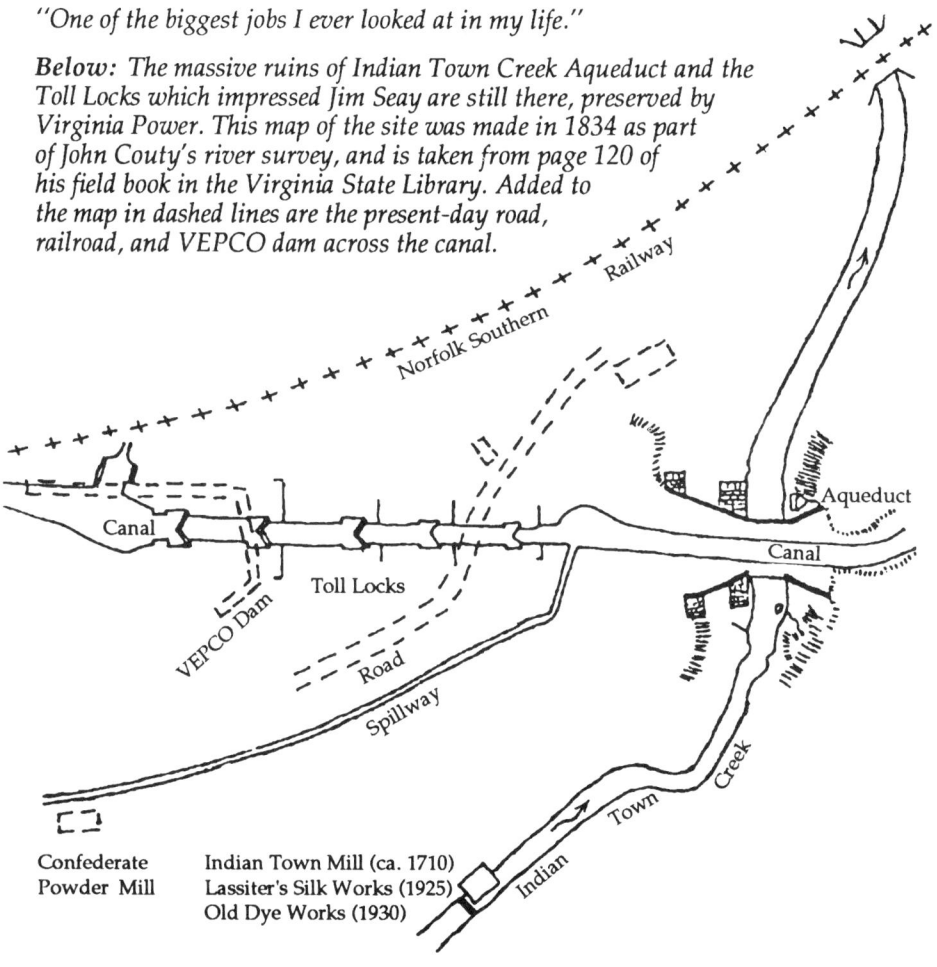

Confederate Powder Mill
Indian Town Mill (ca. 1710)
Lassiter's Silk Works (1925)
Old Dye Works (1930)

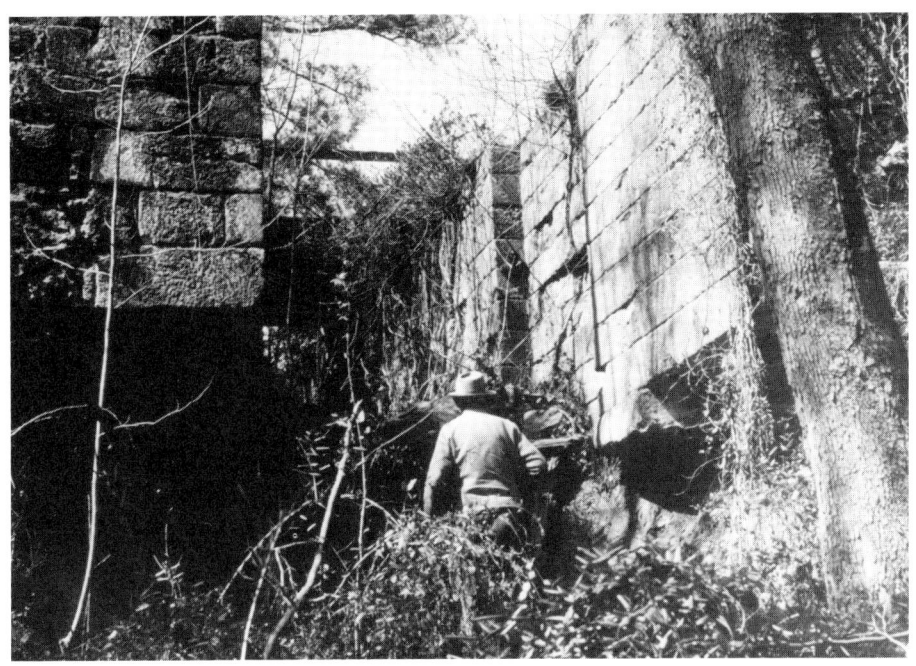

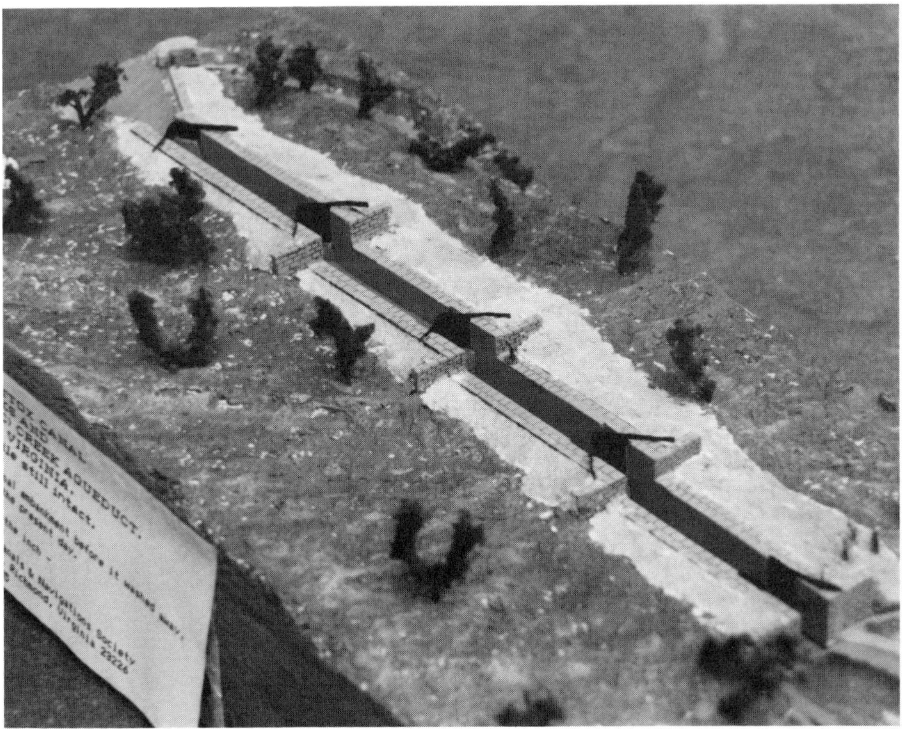
CANAL
CREEK AQUEDUCT,
VIRGINIA.
still intact.
abutment before it washed away.
present day.
Navigation Society
Richmond, Virginia 23226

"I'm going to try to get these people going back and forth across the river here who want to come over to Ferndale Park."

Enlargement of the cover photo, showing Jim Seay's batteau in use as a ferryboat, with a passenger cabin added. The view is up the Appomattox River, with the Matoaca Mills smokestack on the skyline above the cabin. (Postcard courtesy of Lee A. Wallace, Jr., and Fred R. Bell)

CHAPTER 2: JIM SEAY, FERRYMAN

"I took my old boat out the canal over yonder and pulled it down in the river and used it for a ferry and made a lot of money . . . round trip 5 cents."

JIM SEAY'S BATTEAU BECOMES A FERRY BOAT AT MATOACA

"I married in 1900, something like that. Then I was running my old river boat. And after they started work on the big canal and all [for hydro power] and discontinued their boatmen. And I took my old boat out the canal over yonder and pulled it down in the river and used it for a ferry and made a lot of money . . . round trip 5 cents. Nothing but people, nothing but passengers. Sometimes I'd have a hundred at a time going across, especially mornings, people going to work. . . .

"The first Sunday when Ferndale Park was beginning to be patronized right much, and the streetcars ran over within two miles of Ferndale Park, and the people could ride out there in the country to what is known as Central State Hospital. That's as far as the cars came. Then they extended the line on out to Ferndale Park, it began to be a big thing.

"Then the first time I was operating my bus line ferry across the river I was operating with a pole, didn't have no cable across the river, and a friend of mine named Frank Clements operated a place in Petersburg, or two or three places. And he was pretty wealthy. And when he come down by the river where I was, he said, 'Jim, what are you trying to do?' I said, 'I'm going to try to get these people going back and forth across the river here who want to come over to Ferndale Park.' He said, 'You can't handle it like that, can you?' I says, 'No, I expect to put me a cable across the river after a little.' And he stopped a minute and says, 'You know where my place is?' I say, 'Yes, Sir, Mr. Clements, I know where it is, it's known as Palms.' He said, 'Well, you come down there in the morning.'

32 *Appomattox River Seay Stories*

"I didn't know what he wanted, hadn't talked a word about it. Went down there the following Monday morning and he says, 'What do you want towards to fixing your ferry up?' 'Well,' I said, 'I don't need anything except a cable.' I think the cable cost me about $15. A 5/8 cable across the river 250 feet long. He write me a note to Leonard Hardware: 'Let Mr. Seay have anything he wants and charge it to me.' And I didn't have to worry then, starting up my ferry boat. So I put that ferry boat across there."

HOW HE LAID OUT HIS FERRY CABLE

"The river did the work [to move the batteau ferry boat]. But you see, I don't know, I never figured it out when I put my cable across there. My cable must have been exactly square across the current. If it had been the least bit lower down on one side than it was on the other, why you'd been going uphill going one way.... I planned it to be square and straight across, [using] nothing in the world but my eye.... Worked for years and years. I used it when the water was so high I put my artificial walkway out there on the bridge and loaded people on the side of the boat.... There was a walkway right in the boat, and in two seconds you're across the river. I can hear the old pulley wheel whistling now, got it hanging up down on the house."

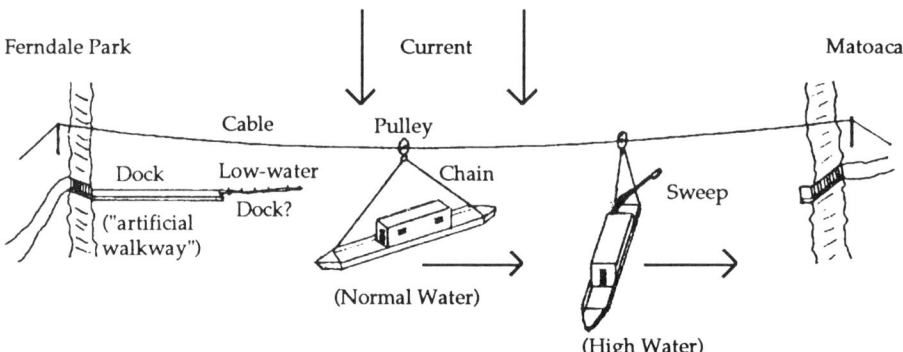

"The river did the work."

Hypothetical diagram of Jim Seay's batteau ferry operations, based on the two postcard views and his reminiscences. Thanks to Mr. Clements of the Palms Ice Cream Company, he obtained a cable to stretch across the river. Chains from a pulley to the boat were let out or in to place the boat at an angle to the river current, which pushed the boat across. When the water was high and swift the boat was put end-on to the current. Even after the ferry was replaced by a bridge, when the bridge flooded Seay put his boat and "artificial walkway" on the bridge to let people across. Whatever happened to Seay's ferryboat pulley? (Drawing by W. E. Trout, III)

HE INVENTS A PERPETUAL MOTION BATTEAU FERRY

"And let me tell you, I learned something that probably might be worth something to me today as I practice. You could call it perpetual motion. And that's never been . . . accomplished. Let me tell you what I saw. At the Jamestown Exposition I saw a wheel that was supposed to be as near to that power as was ever known. It had five spokes and its spokes were hollow. And each spoke had a ball in it, a solid iron ball. And that ball rolled from the rim to the hub as the wheel revolved. And you give it a turn and when this spoke gets to going up, it's toting up one [ball over to] the other side [so] there's three going down, there're five, you see. The odd spoke, that's the one supposed to take care of the uneven weight on the wheel. That's the way to arrange it. But it never did work. You give it a turn and it runs for awhile but finally all of the balls would get on the low side. But it would go for a long time. . . .

"I was going to tell you about the boat now. In low water I operated the boat lengthwise across the river. The boat was 60 feet long, had a chain from here, [and] here [probably attached near each end of the boat], and up to a pulley on a cable. And I poled it with a pole. Well, when the water got extremely high we'd have to turn it around and put it end way to the water. . . . Go across sideways with a chain hooked on each side, same distance from the head of the boat down on each side. And if you shorten *this* chain to let *that* chain out to turn the boat [at an angle] to the stream, right across the river it'd go. . . . When you turn the boat in that angle like that, the water running against the side of the boat would carry it right on across the river.

"Had a sweep on the boat, which is a big paddle, on the head end of the boat. That was in the water. If you just turn that sweep the least bit and fasten it to this side of the boat, that boat would go across the river so fast, you would have to keep it from bumping too hard on the other side. All you had to do when you got over there was to tighten up on *this* chain, and slack off on *this* chain, and turn the boat in that position, and right back across the river it would come. The current did it. . . .

"Now, I could arrange the thing, on the bank over there, that when this boat comes over there and bumps that thing, it bumps that oar back in this shape, and it comes back by itself! When it comes back on this side of the river there's something in the water here of that description [to] raise the oar and shove [the boat] around, to go back itself. And if that wouldn't be perpetual, I don't know what would! . . .

"I thought about that thing a lot of times. I could arrange the thing over there, talking about that perpetual, so when the boat bumped the shoreline, a jigger would drop down and hold the boat. If I wanted to hold it five minutes, I'd arrange the jigger to turn the boat loose in five minutes. And the paddle had already been rechained and reset to go back across as soon as that release, that catch on the boat, would go back across. Same thing on the other side. And I always thought that might be perpetual motion."

"We used to have some great fishing places up this river."

On his exploratory trip down the Appomattox two centuries ago, Benjamin Henry Latrobe sketched these two fishermen about two miles above Clement Town. Could that be a keg of whiskey in the boat? (From the Spring 1959 **Virginia Cavalcade**, courtesy of the Virginia State Library and Archives)

Unfortunately for posterity, Latrobe fell in the mud on his way to Petersburg and ruined all his drawing paper. Thus he had no way to record, with his watercolors, what he saw in Petersburg.

CHAPTER 3: JIM SEAY, OUTDOORSMAN

"I have had more experience, I reckon, than any man ever had, because I've always been an outdoor man and I never worked in a roof in my life over thirty days."

ON THE OUTDOOR LIFE

"I have had more experience, I reckon, than any man ever had, because I've always been an outdoor man and I never worked in a roof in my life over thirty days. I worked in the Old Virginia trunk factory down here, took up concrete work when I was a young man, and I built a concrete foundation for a 300-horse engine and the thing took me around thirty days and I was working under a roof, and that's the longest I ever worked under a roof in my life.

"I worked outdoors all my life. I don't know if I can attribute that to my long life or not and I never took care of myself to no great extent. I've hunted all my life and slept in the woods, slept on the wet ground. And my hip bone made a hole in the ground and there'd be water the next morning. Now that shows you I didn't take care of myself!"

JIM SEAY GOES FISHING WITH MARY LOUISE

From a 1990 Interview with Hubert Holt

"Jim Seay was crazy about my stepmother. After her husband [my daddy] died, he used to court her. They were both river rats, they would go up the river and fish. You know where Allen's Marina is located, that was Namozine Creek. She told me they went up there in January, it was turning real warm. He had an old Model-T Ford.... They went up there to go fishing. After you cross that bridge, if it's wet weather you don't dare get off the road, or you get stuck.

"They walked down the side of the creek about a mile to the mouth, where it dumped into the Appomattox River. I know exactly where it is now, even though it's all covered with water.... Mr. Seay and Mary Louise, they went down there and fished and they were having such good luck that they didn't want to leave, so they stayed until they just barely had time to walk out of there before dark, carrying their fish in a sack bag. They got the old car started and they tried to turn it around, a narrow road there, and his rear wheels got off and went to spinning.

"Well, Sir, she said, 'We worked and worked and worked trying to get that thing out, and it went deeper and deeper.' So finally, about 10 o'clock at night, they gave up. They decided to walk home. She said it started snowing. And she said they took off and they got in about 4 o'clock the next morning, they walked all the way.... That's a long ways, you know that? That woman was tough, I'm telling you!"

ESCAPING THE ICE JAM

From a 1989 Interview with Frank Seay

"Did my Dad ever tell you about the time he was coming down the river and they had this ice jam? He'd come through [Goode's] Falls, and he was headed for [Goode's] dam, and all at once the water started rising. Cold winter time. The ice had just broken up on the river. So he knew he couldn't go down where the ice jam was. And the water got so high it went out in the woods. So he left the river bed and he went [and paddled his boat] into the woods 50 feet or more to get away from that ice jam. And all at once that ice jam broke and left him out in the woods! So, what he had to do, he had to take out his axe and his saw and cut trees and put them under his boat to roll it back to the river. He always carried a block and tackle with him so he could hook onto something to pull that boat."

Jim Seay's homemade paddle, on display in Olger's Store, in Sutherland, in 1981. (Trout photo.)

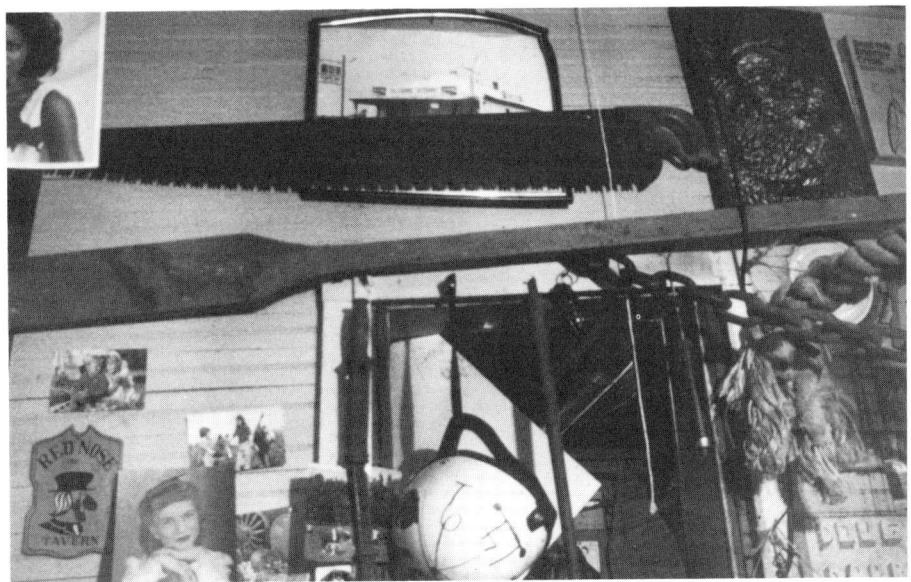

BLACK'S CANAL AND MR. WEBBER'S CARVED STONE

"You know, right where I've got my shack up yonder on the river . . . there was one of the biggest jobs that they had on this whole line of river work. They had to dig this canal, Black's Canal, . . . about a mile long. And the extreme upper end of it consists of a big rocky hill they had to cut through for that canal and required I reckon more blasting and drilling than it did on any other place on the river. And I understand that . . . the overseer of the work was named Webber. And my boy [Frank] found a rock in the river up there about as big as a plate or larger with a 'W' cut on it, and I told him, I said, 'Save that, I bet old man Webber put that name on it, that initial on it, just W.' And his name was Webber. And while he sitting there probably watching them drill he chiseled that there on the rock. Found right down in the canal. It got misplaced, I don't know what's become of it. It would be covered up now with water when that pond fill up, I'll never see any more of that."

MR. WEBBER'S STONE (Part 2)

From a 1989 Interview with Frank Seay

"My Daddy had a cabin on the island and I used to spend a lot of time up there. And that was right at the end of [Black's] canal. One day I found a stone with a 'W' on it, the thing was three or four inches high, the lettering. Flat stone. And I took it over there and showed it to Papa, and I said, 'Papa, what's this W for?' He said, 'Well, I don't know unless it was put there when they dug the canal, because there was a man in charge of this work named Webber.' And it could have been somebody carved that W on there and put it beneath a big pine tree, right in the roots of the tree and it was there when I found it."

THE STRANGE ROCKS FROM THE POTHOLE

"Here's a strange rock.... Look at that one.... I've got a lady friend that cooks biscuits, I told her that was one of her biscuits! ... One of the things that I see that makes rocks round in the bottom of the river: In the bottom of the river in a big solid rock large as this floor and level as this floor, you will find a round hole big as a pot. And in that hole you will find most any beautiful shaped rock you want to look at. And the water causes the rock to go round and wears a hole out in the rock."

"Here's a strange rock."

Pocahontas Basin, the most famous Appomattox River pothole (shown here in its previous location beside the old Pocahontas Bridge in 1906) is now in Poplar Lawn Park. (Fred Bell Collection)

JIM SEAY FINDS THE BODY

From a 1990 Interview with Philip R. Roper, Jr.

"Another thing about Jim, if anybody drowned in the city of Petersburg, and they couldn't find the body they always went and got Jim. Now why they didn't go to Jim first, I'll never know. Jim Seay was always available. He never let you pay for it though.

"But I had a young girl up here that I was going with then, she was named Mary Harrison. Her brother and a friend of his, Holmes Boisseau, went hunting and canoeing, went up there up near Farmville and they started down that river and they never saw them again. But somebody found the canoe so people went out there and they searched that river from a to z and they never did find the bodies. Somebody said, 'You want to find them, go get Jim Seay.' So Jim went out there and he said, 'Now where was the canoe?' And they said, 'Right here.' So he stayed around there for several hours. Looked at the current of the water and so forth and finally he went down about four or five hundred yards from where the boat was and there were these two bodies, lying on the shore. Jim was a master. Had a man drown on Wilcox Lake, couldn't find him. They finally got Jim, Jim went out there and found him. If you ever wanted to have anything done around the water, Jim Seay was the man to get.

"Jim Seay found a rock, a big boulder, up there near where those boys were found, George Harrison, Jr., and Holmes Boisseau, Jr. And that rock was brought to Petersburg and put over here in the Christ and Grace Church yard on Sycamore Street. Mr. Seay found that rock, or so I've been told. It's a great big thing. Somebody cut the names of those boys in that piece of stone."

Monument to George Harrison, Jr., and Holmes Boisseau, Jr., at Christ and Grace Episcopal Church, 1545 South Sycamore Street, Petersburg. Frank Seay was also part of the search party and recalled that the accident took place on Christmas Eve in 1932, and that Jim Seay had discovered where the boys' clothes had washed up, after they took them off to swim to shore. Boy Scouts finally found the two in the low grounds, frozen to death. (Trout photo)

JIM SEAY'S LOST LOG CANOE

"Let me show you my old Indian canoe. . . . I had a real nice looking one here . . . sold it to the county . . . they got it in Chesterfield Court House, they have a museum there. . . . My boy has one in his yard but his is not in as good shape. . . .

"Now right where my little cabin is I've got up the river there's one laying in the water just like that. I can feel it but never have seen it. . . . Been under the water ever since I ever found it. Found it at extreme low water when it's clear and I saw it down in the water. And I can put my hand on it in low water. Right now there's about eight foot of water over it. . . . It's 100 feet downstream [from my cabin]. . . . If I could have been up there when that man was up there on that island with his machine [clearing for Lake Chesdin] it would have been a simple matter to pull it out."

"Let me show you my old Indian canoe."

This dugout canoe, one of several discovered by Jim Seay, is now on display in the Chesterfield County Museum. (T.T. Brady photo)

HOW THE HISTORICAL SOCIETY GOT ITS DUGOUT CANOE

From a Report by Olin L. Taylor
(Courtesy of the Chesterfield Historical Society)

"In the Spring of 1969, Mr. Fred Pease, Chairman of the Museum Committee, called me and informed me that he had information that an old wooden canoe had been recovered from the Appomattox River and was at the home of Mr. Jim Seay on River Road, Matoaca. Immediately, Mr. Pease and I went down to investigate the report of the canoe. Mr. Seay most graciously welcomed us and showed us the canoe that was stored in a long shed on his property. We asked Mr. Seay how did he find the canoe that he said was almost completely buried in the mud in the bottom of the river. He said he was walking along the river bank and noticed a distance out a circular movement of the water in a spot he hadn't noticed before and he thought there must be a stationary object beneath the water. Finally his curiosity got the best of him and he went out in the river to investigate and found the end of the canoe protruding enough to tell it was not just the end of an old rotten log.

"Mr. Seay said it was quite an undertaking but he and his sons finally got it out of the river and moved it to his place. We told him we thought the Chesterfield Museum Committee would be interested in restoring it and putting it on display at the courthouse with the other Indian artifacts we have collected and would he consider donating it to the museum. He said no as he and his sons planned to restore it and show it along with their collection of Indian relics that they had collected in the area over a period of years. Soon afterwards we went down to talk with Mr. Seay again about the canoe and to look at his collection of Indian relics. Again Mr. Seay said no he had talked with his sons and they said they would not part with the canoe.

"I can't recall how many times we went back to see Mr. Seay but they were numerous as Mr. Pease and I were desperate for the canoe for the museum. The last time we went back we started talking about the canoe and told Mr. Seay we felt it was a historical item of Chesterfield County and we would like so much to have it for the museum and we were so anxious for it that we had $50.00 we would pay him for it.

"We were two happy people when Mr. Seay said, 'You fellows are so sincere about the thing I will sell it to you.'

"I still don't know whether it was our sincerity or that Mr. Seay was just tired of looking at us."

"He called it 'Gould's Mansion.'"

Only these earth mounds remain of Gould's unfinished mansion, also known as "Rock Castle." Gould's high dam was never begun, but engineers surveyed the site, using Jim Seay's houseboat as their cookhouse. Frank Jay Gould, son of the famous Erie Railroad financier Jay Gould, was first vice-president and director of the Virginia Passenger & Power Co., which owned the Upper Appomattox Canal and later became VEPCO. Gould's Appomattox River project deserves some research. Another version of the story is that the mansion was begun by Mr. Andrews who sold it to Gould, who in turn found it could not meet the building codes. Work on the dam is said to have been stopped by the government as an anti-monopoly action. (Trout photo, 1990)

CHAPTER 4: JIM SEAY'S HOUSEBOATS

"He spent a lot of time on that river, in that houseboat" (Frank Seay)

A DAM WHICH WAS NEVER BEGUN AND A MANSION WHICH WAS NEVER FINISHED

"In nineteen hundred and four, I think it was, the Goulds were going to build a big dam up there just like they're building now [for Lake Chesdin]. They contemplated building that sixty feet high. And I was on the river trapping and hunting at the time in a houseboat. And the engineers, surveyors and all was up there and I was feeding them out of our house boat, had a cook, the cook wanted to move the boat in where they wanted to go to build that dam....

"At Goode's Bridge, Jay Gould, he was going to have that as his own—he called it 'Gould's Mansion.' The greatest game place you ever saw in this life. Turkey, deer, beaver, and everything else was running wild. So he located that and he was going to have a home built there and he built the house out of rock picked up around the field and the building is the shape of a clover leaf, three round rooms and a hall representing the porch, the stem, of the clover leaf. And each room had a fire place in the same chimney, on the way up, one chimney accommodating each room with a big fireplace, and had wooden steps going from one floor to the other, three floors, three stories high including the basement.

"And after they discontinued with the dam, Gould threw it up and didn't have anything else to do with it. Had never finished, had never completely finished it. And a man named Andrews bought it. And Andrews went to the place and he found part of the steps. He says, 'I've always had some kind of a feeling that I was going to be burnt up someday and I'd rather you take those wooden steps down and put the cement steps up, to be fireproof.' He goes back to his home in New York and spent the night in the Astoria Hotel and in a big fire burnt up!"

HOW JIM SEAY RAN HIS LEVEL

From a 1989 Interview with Frank Seay

"When we had that cabin on the island, which was a houseboat added on to, because the bottom got leaking, so he put it on the island and built a room onto it. This room was more or less for sleeping quarters, and we cooked in the houseboat.

"So the houseboat got old and the other part of the building was old and he built a new cabin. This one was built up high and I went out in the woods and helped him get the cedar posts to put on the island to set the house on because we wanted to raise it up. The new one had to be higher than the old one because we had trouble lots of times with high water getting in the cabin.

"So I said, 'Papa, where's your level?' 'Don't need no level, boy, on an island.' Well the island was about as wide as from here to my car there, a little more [about 30 feet], so he goes to the edge of the island and puts down one post, he goes to the other end and puts down another post, then he runs his line through there and measured down to the water, and got his level that way, and that's the way he sawed the piling off! So it was level.

"And I guess it was at least five feet or close to it above the island, and about four feet in the ground. And we had somewhere around ten or twelve pilings. [When Lake Chesdin started to fill up] all you had to do is put the barrels under it and wait for the lake to raise up and he had it. Wasn't no problem putting the barrels under there, the problem was getting it loose from the pilings. My brother helped him move it. I think they pulled it with an outboard motor."

Jim Seay's Houseboats 47

"I put 24 barrels under it, and now it is floating on the water like a cork."

There are no known photographs of Jim Seay's houseboats, but there are several of his cabin, which he floated off on oil barrels as Lake Chesdin filled up. In this photo, the cabin is floating in the background to the left of Jim Seay's hat. (Trout photo, 1968)

"I think they pulled it with an outboard motor."

Another view of Jim Seay and his cabin floating in Lake Chesdin. When Lake Chesdin was full, he towed the cabin up to dry land at the head of the lake. (Trout photo, 1968)

JIM SEAY'S HOUSEBOAT

From a 1989 Interview with Frank Seay

"[Papa] showed me a picture one day of some men that came to visit him on his houseboat. And it didn't look like the houseboat that he had in later years. And he said that they were at Deep Creek. They were at the mouth of Deep Creek and all these men probably had driven up there during the early days of automobiles and they wanted their picture taken.

"So when they were getting ready for the photographer, they must have had a regular photographer with them, when he was setting up his camera and his tripod some man walked up to the shore line and asked him if he could get in the picture. 'Why sure, come on aboard.' So he went on and got in the picture. And my Daddy said never did know who the man was! I said, 'Papa, let me tell you who it was. There was only one family living in walking distance of there and his name was Poindexter. There was a Poindexter family living there, he was a farmer, and that's who it was!'"

HOW THE PREACHER'S SON FELL IN THE RIVER

From a 1989 Interview with Frank Seay

"Let me tell you a joke. This Mr. Bunting, the preacher's son, a salesman, he was sitting on the stern of the [house]boat while Papa was pulling it from the bow [with an outboard motor on a rowboat]. You could use it either way, it was the same on both ends, had a deck on each end. Anyway, Mr. Bunting was sitting on the back deck in one of these folding chairs, and he always did this, [leaning back], you know, and he sat there, he was backed up to the river. And he had a little white dog, a squirrel dog, more or less, and that little dog loved him and he went and jumped in his lap, and overboard he went! He got out of the river fast, it was cold water!

"That was up there near Black's Dam. The [house]boat he used mostly, the one he pulled out on the island after it started leaking, was the boat he built in the yard here at the old home site. It was built in 1914. He and a black man cooked the meals for the surveyors [for Jay Gould's proposed dam] and that's where they ate."

"Did you know that people came out there in such a quantity that cars ran every ten minutes?"

Trolley Car 610 waiting at the station in Ferndale Park for departure time. A gum machine on one of the columns dispensed one-cent packs of gum. If you got a pack of a certain green color, you got another pack free. From the collection of M. D. McCarter, in **Rails in Richmond**, by Carlton N. McKenney (Interurban Press, 1986), courtesy of Mr. McKenney.

CHAPTER 5: FERNDALE PARK

"Ferndale Park was a beautiful place."

FERNDALE PARK IN ITS HEYDAY AND WHAT HAPPENED TO THE LITTLE DOG

"They didn't have anything out there but a few tables and swingers and they had balloon ascensions to attract people to go out. A lady was going up and she failed to show up, I believe, and they decided to send a dog up in a parachute. And they fixed a little harness around the dog and put him to swing from the parachute, but I don't know how he was going to arrange to cut himself loose. But anyhow, after the dog got up about 40 or 50 feet high he wiggled out of the harness and fell out there in a swamp. If it hadn't been mud and water up to your waist, could have killed him. Fell in that soft mud and didn't hurt him. And I went with the boys out there to get him out of there. And they had swingers, half a dozen people swing at one time in a box swinger. That was all the attraction they had.

"Then the power company took that thing over and extended [a streetcar line from Petersburg] to the Norfolk and Western Railroad. That was within a half a mile of Ferndale Park. It then ran out there for some time and people walked from here over there to catch the car to Petersburg. Then they brought it over to the canal, the electric cars, tunneled under the Norfolk and Western Railroad and ran their cars over here [to Ferndale Park].

"Then they discontinued with the old river boats in the river and I took my riverboat out and dragged it down to the river and used it for a ferry to carry the people back and forth across the river. And done well with it, and ran it until they built a bridge across the river. I used to make sometimes 40 or 50 dollars a day when they had big picnic times.

52   Appomattox River Seay Stories

"Ferndale Park was a beautiful place. They had a moving picture house there, the first moving pictures ever shown around Petersburg were shown there. No talking. Pat Norlington was a young boy about my age, and while the pictures would be showing, he would be behind a curtain making a noise. Pat Norlington was his name. And he had a piece of tin, thin tin, and if the picture was showing a wind storm or the wind blowing he would shake the tin to make a sound just like wind blowing. And when the horses was racing he had two English walnut shells and he would tap them shells on a pan or something, sound just like a horse's feet. No talking, just moving pictures. They were the first.

"Did you know that people came out there in such a quantity that cars ran every ten minutes? And sometimes they'd have to run 'til one or two o'clock in the night, when 12 o'clock was deadline, to get the people back to town, there were so many of them! That was Ferndale Park!"

"They had this wading pool . . . between the canal and the river, that's what you have the picture of."

Postcard view of the park's fountain, probably taken at the same time as the cover photo. Jim Seay's batteau ferryboat is in the river just beyond the peak of the fountain. (Courtesy of Fred R. Bell)

FERNDALE PARK (Part 2)

From a 1989 Interview with Frank Seay

"I spent a lot of time over there at Ferndale Park. Here's what they had at Ferndale Park. They had the hobby horses, as we called them, [on the carousel,] but most of the amusement section was on the south side of the canal. They had this wading pool down at the foot of the canal, between the canal and the river, that's what you have the picture of. They had a dance hall there in later years, but the dance hall was put in after the park had been there a long time. They had a theater there at one time. A very tall theater. And they showed these old hand-cranked movies there, flickers I guess you'd call them. And it was a huge thing. It was as high as any part of the [Matoaca Elementary] school over there. And we used to climb all up in there chasing those bats. It's a wonder we hadn't got hurt. But they had a ladder on the side of a column going up to the projection platform, it was all open. Didn't have no projection room, just had a projector. And they showed these old movies there. And you can still see where it was.

"They had a Tunnel of Love. You went in a boat and went up and come back, all in the open. They called it a Tunnel of Love but it was a little canal which ran up between the canal and the river. I guess they had to row themselves. That was up from the theater, some distance up from the fountain, but closer to the canal. The fountain was down in the low grounds. It could be covered today by that new bridge. And I believe that wading pool was fed by a spring in the canal bank. There were springs along the canal bank, not leakage from the canal but springs, and I believe they fed that pool.

"There was some pond there in the old days but I don't know that it was as big as it is now. And right there where that nice building is standing, that was the ice-cream parlor. And we all used to like it because they had these metal tables, metal tables and metal chairs, they were for adults and children. And they had a Victrola, we'd wind the Victrola to play the records. Well that was in the back, that was next to the lake. And the other was out front, that's where you got your root beer and ice cream.

"Well, right from that was a place that you could throw baseballs, concession stands, win candy and so forth. And then they had a shooting gallery there, used a .22 rifle. And then between the shooting gallery and the bowling alley were swings. And then the bowling alley. And it had a couple of pool tables at one end. And they had about four lanes there. That was on the south side of the canal. And the bowling alley was so close to the canal that sometimes when they had these panels open, a ten pin or a bowling ball would go out and go in the canal! We would set up pins. Five cents for setting up pins. We set up pins until we got enough money we could bowl, don't you know. Didn't know what a pin setter was in those days, except it was a boy!

"[Ferndale Park] was operating up until 1928. It was mostly closed a little before 1928 but there was only the store, and of course the store was where the street car come in, sort of a depot like. And it had a shelter in it where you could sit on the benches and wait for your car. They did sometimes open it up on weekends, had the hobby horses and the bowling alley. So that's the way it was."

"It was mostly closed a little before 1928 but there was only the store . . . ."

The store and pool hall building was next to the trolley tracks in Ferndale Park. The store seems to have lasted until the park was entirely closed down, in 1928.

"They showed those old hand-cranked movies there, . . . they had a ladder on the side of a column going up to the projection platform."

This snapshot of Ferndale Park's theater shows the ladder up the front of the building. The theater was on the north side of the Upper Appomattox Canal, visible in the foreground. A loft for stage-show scenery can be seen in the background.

These seven insurance photographs of Ferndale Park's buildings are from the collection of Carlton N. McKenney. Mrs. H. L. Dunnavant of 7563 Sambar Street, Chesterfield, is collecting reminiscences of Ferndale Park and would like to hear from anyone who remembers it.

Ferndale Park 55

"And they had a shooting gallery there, used a .22 rifle"

Frank Seay recalled that one of the patrons was killed there, in an "unloaded gun" accident.

"They had the hobby horses, as we called them."

The carousel or merry-go-round was always called the "hobby-horses."

"Sometimes a ten-pin or a bowling ball would go in the canal."

The bowling alley, with the swings in the foreground, was between the trolley tracks and the canal.

The only original building still standing in Ferndale Park is the park superintendent's residence, since remodeled and now a private home.

"We'd wind the Victrola to play the records."

The back of the restaurant or ice-cream parlor overlooked the water.

Opposite page:

"Do you know why they call it the Abutment Dam?"

Map of the Abutment Dam, at the head of the Upper Appomattox Canal through Ferndale Park. The original 1800 deed from Robert Atkinson to the Upper Appomattox Company refers to "an island named Sycamore, against which two dams abut." From a 1904 Virginia Passenger and Power Company (now Virginia Power) map, courtesy of the Appomattox River Water Authority.

THE DAY THE CANAL BLEW OUT AT FERNDALE PARK

From a 1989 Interview with Frank Seay

"I'll tell you what that hole was [on the other side of Ferndale Park Lake beside the canal]. That was where they [got dirt to] fill in the canal when they washed it all out. See, it took everything out of there, took the canal bank out too. I would say it was about 1920. And when that thing exploded I could hear it at home. I believe somebody blasted it. I don't know why. But it was a stone culvert, underneath the canal. And when that thing went out it sounded like a stick of dynamite went off, or maybe more than a stick of dynamite, because I heard it at home. They got that dirt out of there to fill in the canal bank, that's what those holes are for. I fished right there, there's a fishing hole there, also a beaver lodge."

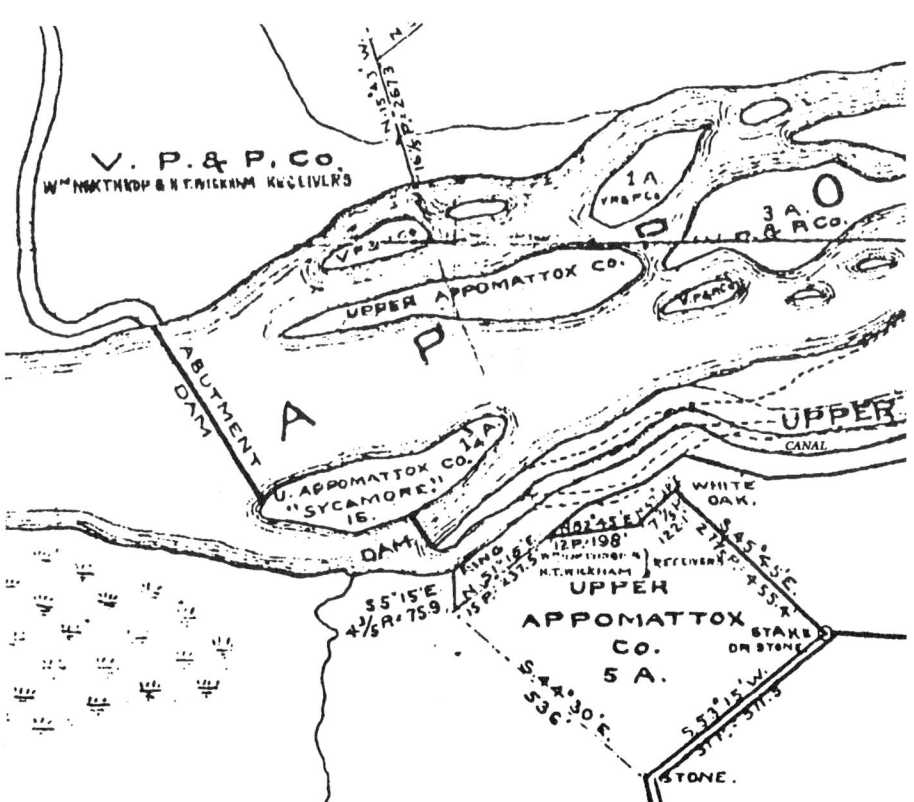

HOW VEPCO WIDENED THE CANAL WITH STEAM DRILLS

From a 1989 Interview with Frank Seay

"Do you know why they call it the Abutment Dam? They built two dams there, one abutted the other. In front of this concrete dam which is there today is a rock dam, right at it, they join together. You can step off the concrete dam onto the rock dam. That was the old original dam for the boats.

"Well, when they [VEPCO] wanted more water to go down the canal, they built the dam higher. And they blasted out a lot of that canal that hadn't been blasted out before. Well, there're boulders laying on the river side of that canal that this table wouldn't touch. You've seen them. You can't walk among them. I walked all through the low grounds looking at those big boulders.

"How did they put them over there? They had steam. And you know how they got steam for their drills? They had a steam boiler they carried along on a track. And they had so many feet of track. And as they brought the boiler down the river, on the south side of the canal, they brought it down along the top of that hill overlooking the canal. And they laid the track as they went along, so they didn't need but so much track. And they used coal to fire the boiler. That wasn't on the river side of the canal, that was on the other side. Up on top of the hill. Well you see they did most of their drilling there. And you can see the drill marks on the other side from the river. That's why they had that boiler over there. I've seen the piles of coal there where they fired them.

"They did a lot of work there. Did you know, they had barges up there for hauling concrete? They had barges up there with wheels on them, at the Abutment Dam. And they would load their cement on the barges and take them down a track and float them to the work site. And some of the barges are sunk up there now, in the pond now, close to where you come through on your canoes. I've seen the water low and seen them in there. They're in there today.

"And they had an uncanny way of fixing a leak in the dam. If the dam got a leak in it, they'd go there and they would drive a pipe in there. The pipe would accommodate the leak. They put concrete all around that pipe, seal that pipe right up, let the water run through it. When the cement hardened, they went in there and drove a wooden peg in the pipe and had the leak fixed!"

Ferndale Park 59

"They blasted out a lot of that canal that hadn't been blasted out before."

The bow of the canoe points to a blasting hole drilled by VEPCO when the Upper Appomattox Canal was widened with steam drills after 1902. (Trout photo, 1988)

"Last summer I believe all the water in this river would run through a barrel at one time."

Aerial view of Brasfield Dam (left) and the lower end of Lake Chesdin looking west. (1987 photo courtesy of the Appomattox River Water Authority)

CHAPTER 6: LAKE CHESDIN

"One of the biggest jobs I ever saw tackled."

HOW THE BED OF LAKE CHESDIN WAS CLEARED OUT

"They have machinery that's unbelievable, I'm telling you the truth. I saw them machines go across that river, through the rough borders, through that deep water, right on across that river just like—well, a muskrat couldn't have gone across nothin like it. Over the rocks, over the islands. And one island out there in the river consists of I reckon several hundred trees, the biggest trees that ever grew on the Appomattox River. Sycamore Trees. Nobody never cut them for no purpose in the world. And those trees were four and five feet on the stump.

"The machine went over there and shoved those trees down out of the ground and buried them and burned them together. Out in the middle of the river. I told the man then, I said, 'I've seen the water here ten, twelve foot high and if you leave your machine out there tonight you're liable to lose it.' But he never did have no trouble. Had a strike of low water . . . just luck. Two years, never had a high water. Generally have three or four every winter . . . . Last summer I believe all the water in this river would run through a barrel at one time."

"We had to paddle like the devil to make it in a day."

Col. Llewellyn F. Rose, the first manager of the Appomattox River Water Authority, took this photo on a trip down the river on September 21, 1965, just before Brasfield Dam was built. Jim Seay, center, is in the stern behind Water Authority board member Orval Hand (left) and Chairman George Brasfield. (Photo courtesy of R. D. Hartman of the Appomattox River Water Authority)

HOW COLONEL ROSE GOT HIS TAIL WET

[Looking at a photo taken on Sept. 21, 1965]: "Here we are when we went on the river. Here's the gang who went. Here's Brasfield, and myself, and Rose, and Hand. . . . And here's a card that Brasfield sent me after he got back home. . . . We went about 30 miles, in other words we went to the extreme end of where [Chesdin] lake's going to be. We put in a small boat and came all the way back down to the dam site. . . . We happened to catch a very extreme low water that hardly accommodated the boat, four of us in the boat, and we had to paddle like the devil to make it in a day. If it had been a high water or an ordinary water we'd have come right along there in a hurry. And coming on down, Brasfield said, 'Jim,' he says, 'do you reckon we'll make it down by dark?' I said, 'Depends on how you paddle!' . . .

"And Mr. Rose, is standing back here now out of sight, he's taking the picture. . . . I'm pointing across the river to where there used to be a bridge. I'm standing with a straw hat on. And Brasfield's down there by me. But Rose is standing back taking a picture. And in less than one minute, he had that picture! What do you know about that! In less than one minute, . . . fifty-some seconds, that picture would turn out. . . . Well, we slipped the boat right down that bank and pulled out for the dam. . . .

"But I was going to tell you about Rose. Rose, he wanted to sit on a seat alone in the boat so he could look at his maps, have room to open his maps. . . . Brasfield and Hand set on the middle seat of the boat and I set on the back seat, I was steering the boat and paddling it . . . . And Rose said, 'I want to sit on the front seat.' . . . He had to sit with his back forward in order to sit there, going backwards all the time. So we had to go over a bad place of rapids in the river, and when the boat went down over that, old Rose got his tail wet!"

"He didn't answer any questions."

Jim Seay blasted this hole into "Lafayette's Rock" for a mysterious man from New Hampshire. (Trout photo, 1976)

CHAPTER 7: JIM SEAY AT WORK

"I've been using dynamite all my life . . . . Now, you could talk with all the dynamite men, with the government, and everybody else, but they ain't got as much sense about it as I got."

BLASTING THE MYSTERY HOLE IN LAFAYETTE'S ROCK

[Edited from "A LaFayette Legend of the Petersburg Area," reprinted from the 1951 (?) *Richmond Times-Dispatch* in *Blandford Lodge No.3: A Bicentennial History*, by W. M. Brown, Plummer Printing Co., Petersburg, 1957. T. W. Haxall Collection, Valentine Museum, Richmond.]

"The tall, aristocratic gentleman from New Hampshire didn't want anyone to know why he drilled the holes," explained Matoaca postmaster C. Clagett Wells. "He asked me if I knew where he could find the rock with 'LaFayette' carved in it. I'd never even heard of it.

"So then the man asked me where the old Randolph mansion was and, of course, I knew that and I took him and his two sons down to the old foundations. When we got there, they took out an old map—it was yellow and cracking—and paced this way and that way 'til we actually found this rock with 'LaFayette' carved on it."

The rock is still there, about 100 yards north of the Appomattox River, on the old Randolph plantation, now a part of Virginia State University. The big hole is drilled in it and the name "LaFayette," almost completely weathered away, can barely be discerned at the base of the rock.

"Soon as we found the rock," Wells said, "the man sent me away. I didn't ask his name because he never asked mine and I only knew he was from New Hampshire because he had a car and that's what the license plates said."

The tall, aristocratic gentleman then set his two sons to drilling, while he sat on top of a near-by knoll, checked samples of the rock and made sure no one came snooping around.

When they had to blast, he hired James W. Seay, a Matoaca builder, now 73 years old.

"He didn't answer any questions," Seay remembered. "Just told me he wanted only four charges so there wouldn't be any chance of cracking the rock. I sneaked in a fifth and he never even knew it—or at least he didn't let on he knew it."

Seay described the hole as "about three-foot high and in about seven feet, then down three feet, then in about seven more feet."

"His sons did all the drilling," Seay said. He had a $4,000 outfit with an air compressor and they operated the rig while he directed. "When I'd done my job, he thanked me, paid me and made it pretty plain I wasn't wanted any more, so I left." The following week so did the tall, aristocratic gentleman from New Hampshire.

Occasionally, during the past 40-odd years, folks in Matoaca have speculated on the reason for the big hole in the grey granite rock.

One school of thought maintains there was a treasure buried under the big rock. But the skeptics insist it would have been easier to dig under the rock than to have bothered with all the complicated drilling and blasting.

Another school believes the hole was drilled for a tomb, but it's still empty today.

Now, with memories getting dimmer, even speculation, with the graven name, "LaFayette," seems about to disappear. There's a small rock near the big one inscribed with the legend: "A. A. 1908." No one knows how it got there. No one in Matoaca, at least, even knows anybody from New Hampshire with those initials.

LAFAYETTE'S ROCK: Part 2

From a 1989 Interview with Frank Seay

"My theory is they dug in the wrong place. I don't see how anyone would expect to find something by drilling into a solid rock. My Daddy asked a park historian if he knew why this doctor was having this work done. And he said they were looking for some of the manuscripts of William Shakespeare! He [the doctor] got his information from Iron Mountain, Michigan. He rented the most modern equipment he could get, which consisted of an air compressor, air hammers, and two young men from here in Matoaca. One of them named Woody Archer, the other Woody Adams. And they worked in there getting this rock out after my Daddy blasted. Well not only that, they worked in there drilling the holes for the blast. And he always told Papa, 'Be careful don't crack the rock.' Which is a huge thing. My Daddy told me that below where the shale is piled against the rock was LaFayette's initials eighteen inches high. That's all covered up."

HOW JIM SEAY BUILT LAKE MARGARET DAM

From a 1990 Interview with Philip R. Roper, Jr.

"I first met Mr. Seay back when I was eleven years old, that's 67 years ago. My stepmother wanted a place on the water and so my father through the company, Roper Brothers Lumber Company, bought this piece of land out there from Mr. Howlett, on Woodpecker Road. And Daddy came back hunting for somebody that could go out there and build that dam for him. And somebody recommended the only man that could build that dam was Mr. Jim Seay. So he went to see Jim Seay. And Jim Seay had an old truck, and he and his son Mac, as they called him then, he was James Malcolm Seay, we called him Mac, they met us up there. Mr. Seay looked at it and said, 'Oh, yeah, I know this, used to be a grist mill here.' He said, 'The old saying goes that during the battle of Seven Pines it rained for seven days and seven nights and this dam broke and washed the old grist mill away.' And he said the stones are still in that creek.

"He said, 'Yes, Mr. Roper, I'll build your dam for you but I want to build it like I want to build it.' And he said, 'Now I want twenty tons of railroad iron. We're going to reinforce that dam so it'll never break.' And he left a piece of railroad iron sticking out of one of the abutments in the back so people could know there was railroad iron in that dam. Every two or three years the state comes by to inspect the dams to see if they're safe, because of the amount of water behind them, they might come down and wash somebody away. And every time they come out there they'll ask the question, who built that dam? The dam doesn't leak—they say it's the strongest dam they've ever seen anywhere. He put the date on there—1922, that's when he finished it.

"All the concrete for the dam was mixed by hand. Jim Seay and Mac mixed it. There wasn't any such thing as ready-mixed concrete. He had a tent, it had sides that dropped down, big enough for two or three people, he and Mac and one son, I forgot which one that was. When he built the dam, he lived in that tent. He cut the trees, he camped out. He didn't want to live in a cottage or anything, he wanted to live near mother earth, as he called it. When he'd get tired of working he'd tramp through the woods to see what he could see. He could tell you where the turkeys were, where the deer lived, everything.

"When Jim was building that dam for us out there, he stayed on the place—he stayed there. Used to go back to Matoaca on weekends. He cooked his own food out there and everything else. Had a little stove. He'd get up in the morning, sunrise, work till sunset. He built us a place out there on the lake for springing boards for diving, he built that. After the lake filled up, he built that boathouse out there for us. You came out there and asked him, though, to take you up the lake or something, he'd stop what he was doing and go up the lake with you!"

68   *Appomattox River Seay Stories*

"I'll build your dam for you but I want to build it like I want to build it."

Jim Seay's dam at Lake Margaret (named after Phil Roper's mother) is just upstream of "Howlett's Old Mill" dam shown on the 1862 Gilmer map of Chesterfield County. (Trout photo, 1990)

HE CATCHES A FISH AND SAVES A SNAKE AT LAKE MARGARET

From a 1990 Interview with Philip R. Roper, Jr.

"One day he came to me, he said 'Come on down here behind the dam, I want to show you something.' And you know me, I loved Jim Seay. I went down there with him. He said, 'You see that? Thing in the creek back here like a circle. That's the old wheel to the grist mill, I found it the other day and I want to show it to you. It's imbedded in that creek.' He said, 'Don't bother with it, let it alone, the water will preserve it and it will never rot anymore than it is, and maybe somebody one day will be interested in it.' So the old wheel to the grist mill is in that creek, and Jim Seay found it. Jim Seay could find anything.

"So he was out there one day, and he said, 'Do you want a fish to take home?' I was just a little bitty kid. I said, 'Yeah, where're you going to get it from, Jim?' 'Wait a minute.' He went on there and reached down in that water with his hand and caught the prettiest fish about so long, that you ever saw in your life. I said, 'Jim, how did you get it?' 'Takes patience, takes patience. I don't need no fishing line, I don't need no hooks, my hands are what I need.' He pulled a fish out of that lake about a foot long, what he called a flatback. He said, 'You take it home, and you cook it, it's perfectly delicious.' And I did.

"Nobody ever got mad with Jim Seay. I never saw him lose his temper. He was out there messing with those rocks out there, big boulders, to put them in the dam. Once in a while he'd turn a rock up, he'd get a hoist out there and hoist it up, there'd be a snake under there, big snake. I said, 'Jim, kill that snake, kill it!' 'Never in this world, nothing wrong with that snake, he ain't poisonous.' I said, 'What you mean, he ain't poisonous, it's a moccasin.' He said, 'That's a water snake, they ain't poisonous.' 'Yeah, you've got to be kidding, I'm going to kill him.' He said, 'Don't you kill my snakes! Only one snake that I know around here that's at all troublesome is a copperhead.' He said he was scared to death of copperheads!"

HOW JIM SEAY REBUILT THE MATOACA MILLS DAM

From a 1989 Interview with Frank Seay

"Did you ever look at the Cotton Mill Dam? We called it Beech Island Dam, because the island is named Beech Island. And that was used for running the Matoaca Cotton Mill. Now, that dam washed out and my Dad built a section onto it back into the island, and it's still holding. But it was originally almost entirely a rock dam, but he built a concrete dam. And he used quartz rock for a conglomerate. Got the quartz out of the side of the hill. And he had a tramway like, that could roll the wheelbarrows on down to the break in the dam. And I reckon that was as far as from here [his home] to the church down there, the first church. And that was in 1911. His name's on the dam, it's over behind the dam, on the backside of the dam, just before you entered the canal. You can get over behind there and walk and you'll see his name up there, 'J.W. Seay, 1911.' Well you see he wasn't a very old man then. He was born in '78 I think."

"My Dad built a section onto it back into the island."

Jim Seay signed his name on the Matoaca Mills Dam. Did he put his name on any of his other work? (Trout photo, 1990)

THE CAR THAT CAME OUT OF THE CANAL

From a 1990 Interview with Hubert Holt

"Mr. Seay was noted for being a waterman. He contracted just about all his life, in concrete work, building dams and things like that. He was famous for it. I'll never forget, there was a Stutz automobile came over to Ferndale Park one night, Friday night or Saturday night, and they parked the thing in the woods just across the road from where you go into Ferndale Park. Parked it in the woods over there. And something happened, the brakes were released on the thing and it went right in the canal.

"The canal was full of water at that time because they used that canal water to run the generators for the power company. There were no trees to stop it and when the brakes released, it just rolled down the hill and went out of sight in that canal. So the first thing they wanted to do was to find somebody that could go in there and get ahold of that car with a rope.... Everybody said, 'Get Mr. Seay, Mr. Seay!' So they went and got him. And he came down there with his old vehicle, old Model-T Ford, and he had all kinds of block and falls and tackle and rope piled up in the back of it. And he didn't do a thing but take that rope and go down in that canal, he had to go down up here, instead of where it went in, because the water was so swift. He had to get down there to grab aholt of that old car.

"Well Sir, I don't know how long he stayed under there but everybody was worried to death, scared that he'd drowned. Seems like he stayed under there 2 or 3 minutes. But he tied that rope on the axle of that old Stutz car. And I helped pull! The whole gang of us pulling on the rope, the block and tackle and pulled that dern car out of the canal. The most ironical thing about it was, I don't know what they used, rags and everything else they could get their hands on, to get the water out of the car, but don't you know the dern thing started?"

"One of the beautifulest homes I ever saw."

Olive Hill (private) was probably built by the Quaker Petersburg tobacco merchant, Roger Atkinson I, about 1774, for his son Roger. The elder Atkinson came to Petersburg about 1750, and built his wealth over a quarter century when Petersburg was the most important tobacco-exporting center in North America. In the 1760s he acquired his home place, Mansfield, opposite Olive Hill on the south side of the Appomattox, and then assembled 1273 acres for Olive Hill. When he died in 1784, he left the plantation to his son, Roger, who lived there until his death in 1829. Due to the younger Atkinson's interest in improvements to the Appomattox, the General Assembly appointed him in 1792 as a trustee to "clear, improve, and extend navigation" on the river. In 1795, he was elected a trustee of the newly-organized Upper Appomattox Company. The "getaway" or escape tunnel mentioned in "The Ballad of Lady Caroline" is now completely sealed up. The kitchen building jacked up by Seay is the right wing in this photo. Photo courtesy of Jeff O'Dell, from his book, **Chesterfield County: Early Architecture and Historic Sites.**

THE ESCAPE TUNNEL AT OLIVE HILL PLANTATION

"Olive Hill? Know all about it. Rebuilt it. Yeah, I done somethin in this world. When I think about what I have done it makes me feel like [something]. Olive Hill was built by a man I think was named Thomas Atkinson. He was a Petersburg man I think, but it's been owned by many since then. It was known as Olive Hill. Don't know as I ever heard where that came from. I'm sure there was no olives there! But anyway, I imagine it was named after some place in England probably. Every old Colonial home in this whole country had its name. Every one.

"Olive Hill was a beautiful home. One of the beautifulest homes I ever saw. And it had a basement, with a fireplace in it, a big fireplace. I rebuilt the fireplace in my time. In fact it was burnt out and I put new fire bricks in the back of the fireplace. And it had iron bars across the windows and it had a wooden lock on the door and I have the key here somewhere now to the lock. An iron key but a wooden lock. And the key shoved a wooden slat into the socket. A piece of hard wood.

"That was on the basement, facing the hall inside the basement. And that's where they punished the prisoners, their slaves, when they disobeyed. And they had a fire place in there to keep them warm, but they were locked in. And the rest of the basement was used by the people.

"And it was said that they had a getaway, same as Robert Lee had over yonder—if the Indians attacked them on the front, they had a go-away down in the basement, that would lead them down to the river. It followed from here about 200 yards, a tunnel under the ground. And the top of that door is just above the floor in the basement now—it had a round curved top. And I've worked all around it."

JIM SEAY RE-DISCOVERS A LOST ART AT OLIVE HILL

[The main house at Olive Hill is on a six-foot-high basement with brick walls. Beside it, the Kofrons, while they owned it, built a kitchen building down at ground level.] "A man name Walls, I think his name was, bought the old place from the Kofrons and wanted that [kitchen] building raised level with the original building, and I undertook the job.

"He requested me to make it look as much like the rest of the work as possible, especially the brickwork. It was beautiful brickwork. And I was not a bricklayer but I could do most any kind of work. So I undertook it. I jacked that house up and got it up to the level I wanted and built a wall under it and tried to duplicate the other wall which was built way back yonder two hundred years ago.

"And I found out my trowel wouldn't do it. And I looked for all the tools in the country. And no tool would imitate a scratch in the mortar section of that thing that I had and I looked at the engraving in the mortar and I said, 'Wonder what that man did use to do that with?' I couldn't figure to save my life what I could do it with. The point of the trowel warn't nothing like it. So in looking around the wall I found an old cut nail that was one of the first nails that was ever made, made in a blacksmith shop. It had a square point on it. I made me a mortar thing and scratched it and that was the thing. That's what they did years ago. That was the thing. Now you can go there and look at it and you can't tell if my work was done first or done last!"

"I tried to duplicate the other wall which was built way back yonder two hundred years ago."

The "lost art" rediscovered by Jim Seay for his brickwork under the kitchen building. (Trout photo, 1990)

WHY HE HAD TO SWIM ACROSS COHOON POND

"[The Norfolk and Western Railway] was the straightest piece of road in the world. I read that several times. From Petersburg to Suffolk, it was as straight as a string. I built that power line, here to Norfolk, steel power line, for Virginia Electric and Power. And it runs about three or four hundred feet south of the railroad all the way. And it crosses Lake Cohoon, a great big lake.... One of the towers was standing in the lake, in the middle of the lake. My job one time was to patrol the line and see about safety devices and things. In order to keep from walking back across the creek on the railroad I pulled my clothes off and put them on my head and swum across the pond past the power line."

OLD SYL BELCHER AND HIS WATERMELON

"An old man named Syl Belcher [lived across the street from me]. Old Syl Belcher was a low-life white man. Didn't nobody like him atall. Never did work. Married a lady had a little money and ran through what she had. I never will forget one day I was in the street and he said, 'Boy, come over here and split me some wood, I've got half a watermelon in yonder you can eat.' I thought there was someting funny, so I went and looked, it was just the rind. He had done et the watermelon!"

"It was called BEADLE'S DUMMY."

George Beadle opened the "Petersburg and Asylum Railway" line from Petersburg to the Granite Quarries near the Central Lunatic Asylum, now Central State Hospital, in 1888. The "dummy" was a dummy locomotive, a small steam engine covered by a wooden body to make it look like one of the passenger cars. The body helped muffle the noise and calm horses meeting the train on the street. In 1901 the line was electrified and extended to Ferndale Park. This silhouette of a "dummy" (not Beadle's) is from Carlton N. McKenney's **Rails in Richmond**, *which has more details on Ferndale Park and its rail line.*

CHAPTER 8: JIM SEAY, HISTORIAN

"All those old-time homes, instead of facing the highway, they face the river."

BEADLE'S DUMMY

"A man named Beadle Lassiter [Seay probably meant Daniel W. Lassiter] owned everything from where Central State Hospital is now, owned all that land, from number 1 to Joe Phillips' home which is a brick house standing opposite the new Pepsi Cola plant now over on Oakdale Avenue—a brick house. When that railroad was built, that was the last house you saw until you got out to the town. Think of that will you?

"That brick house belonged to Joe Phillips. He ran a wood yard down on the canal and I hauled wood down the river to his wood yard, with a riverboat [batteau]. He called it a load. He ran a railroad track from Chapel Street, where the car barn is now. That was his headquarters. This train consisted of a steam engine and one coach. And the engineer was named Willie Wilburn. Lived on High Street two doors from the corner of Canal Street, left hand side going down. Willie Wilburn. He was the engineer on the train. It was called 'Beadle's Dummy.'

"It ran to what was known as Granite Grove. Derives its name from the stone quarries over there. They had two great stone quarries. They done filled them up. I dynamited the stuff two years ago to fill them up. They did away with the ponds which they had down there on account of them being dangerous. I believe one or two patients drowned in it. And they were going to do away with the ponds. They had a great big heavy dirt dam. I had to dynamite through that thing and lay a pipe through there."

STINGY BLACK BILLY

"Three Rowletts [Howletts?] owned all of upper Chesterfield County practically. They owned along six miles running along the river. And they had a brother that bought land over further in Chesterfield County on Hickory Road. And one was named Tom, and one was named Gus, and this one was named Billy. And Billy was a dark-complected man and they called him 'Black Billy.' And they said that he was so stingy that he wouldn't let you cut a switch off his side of the road. And he never was married and lived in a big home, and was buried in a field. And now the grave is lost. A man that just died two or three weeks ago showed me the grave while he was plowing the land around the old home place."

HOW BOSH REGAN LOST HIS ARM

"[In a stone building still left from Matoaca Mills, there's a] great big stone, probably a foot square. And the inscription on there reads, 'Built by Hugh Dooner, 18 hundred and 34 in the year A.D.' That's a portion of the mill, that rock building. The Traylor boy lives in it. And that portion of the building was where the bales of cotton was opened when they were brought here. They went into that building and then it was treated and cleaned.

"And a man named Bosh Regan operated one of the big machines that beat the cotton up and one of the beaters caught his arm and cut his arm off. I never will forget that. And his arm is buried down there under the steps where he lived at. Doctor Bolling fixed it up for him, Dr. Bolling Jones.

"But that was what happened in that building in my knowing. And I think they had one or two fires in that building in the meantime. If there was any kind of a piece of metal in that bale of cotton, that beater would create a spark and that's why they had it separate from the rest of the buildings."

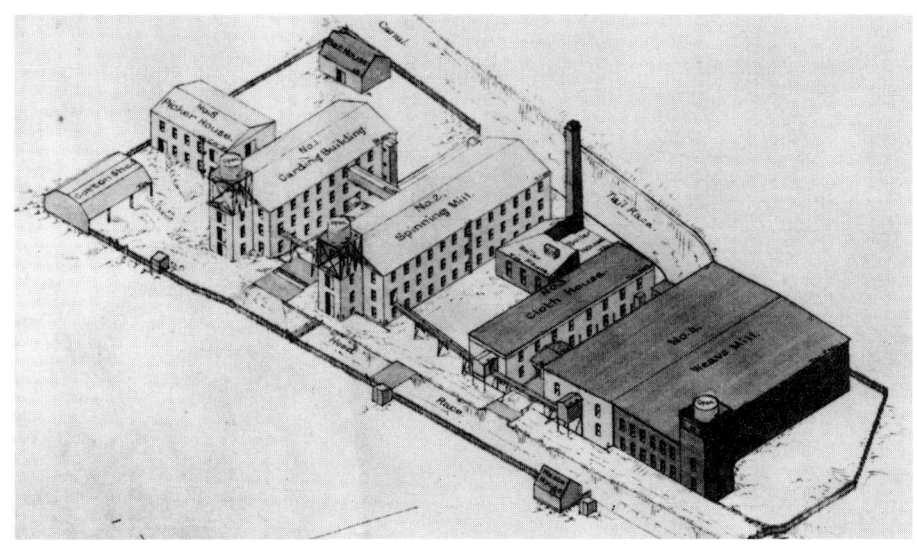

"A man named Bosh Regan operated one of the big machines that beat the cotton up."

Above: *Perspective view of Matoaca Mills about 1902, from an insurance map. Bosh Regan worked in the Picker House, on the left, the only building on this map which is still standing today. Photo courtesy of the Virginia State Library and Archives, from the 1905 Virginia Consolidated Milling Company scrapbook in the Petersburg Public Library.*

Below: *The Picker House, shown here in a picture taken by 1905, is now the home of the Traylors. From the Virginia Consolidated Milling Company scrapbook.*

Top:

The Matoaca Mills Office Building is now a private dwelling across from the Picker House.

Bottom:

An inscribed stone moved from the mill to the Office Building reads, "MATOACA FACTORY/ A.D. 1834/ HUGH DOONER BUILDER." Dooner also developed Dooner's (now Doan's) Alley in Petersburg. Photos courtesy of Jeff O'Dell, from his book, **Chesterfield County: Early Architecture and Historic Sites.**

HOW THE TRUNK FACTORY MADE ITS MOVE

"Mr. Simon Seward and Mr. H. L. Munt, who operated Munt's Mill out at Campbell's Bridge, went in the trunk factory business together, Seward and Munt. And they started to build a trunk factory on the slope of the hill, between the road and the river—and the canal—and Mr. Seward found out there wasn't room enough there for a factory so his night watchman told him one morning, when Mr. Seward came in, he said, 'Mr. Seward, I tell you I had to fight last night to keep this place from burning up,' he said. 'It caught afire in the boiler room and I sure had to work.' And Mr. Seward said, 'Well, next time don't work so hard.' And it burnt up, and he established his plant on High Street."

"Went into the trunk factory business together, Seward and Munt."

Etching of Seward's and Munt's Grist Mill and Trunk Factory at the north end of Campbell's Bridge, from Edward Pollock's **Historical and Industrial Guide to Petersburg, Virginia** (Petersburg, 1884), page 99. (Courtesy Petersburg Public Library)

MILLIONS FOR DEFENCE
BUT NOT ONE CENT FOR TRIBUTE

"Way over on one of the furthest houses that was built from the main building [at Bellvue], was a branch where there was a tremendous big spring. Most wonderful spring you ever saw, gushing water out there a stream big as your arm all the time, been so ever since that house was built and probably that's 200 years old. And right up on the hill from that, what I'm telling you about, the Indians had to have a spring for their village, for their headquarters.

"Well I found lots of relics, just up on the level ground from the spring. And this slave house had been built right there where those Indians had lived. And while I was searching for relics, Indian relics, I noticed the old chimney place and the fragments of the old slave home where it had been torn down.

"And I'll show you what I found in it. Knock you out this time! Down in the fragments of the old fire place I found that. And it reads, 'NOT ONE CENT FOR TRIBUTE BUT MILLIONS FOR DEFENCE, 1841.' . . . It's printed just like a big penny, only its a different wording on it. And I found it in the ruins of that slave house. And a picture of the Capitol in Washington on there and the head is just like an old big penny."

"It's printed just like a big penny."

Seay's coin was probably similar to this one, taken from **Hard Times Tokens** by Russell Rulau, Krause Publications, 1987. Similar tokens substituted for pennies in 1834-1844 when real cents were hoarded and scarce. "Millions for defence, but not one cent for tribute," goes back to 1796 and later to the war against the Barbary Pirates. It was a patriotic slogan which had the decided advantage that the words "NOT ONE CENT" could be used to make it clear to the government that the tokens were not illegal coins. (Photo courtesy of Russell Rulau)

WHY DICK JONES WENT TO JAIL AFTER HE DIED

"The man that built Bellvue, old man Dick Jones, I think he was—he was very old—and they had a law at that time that if a man died owing a debt, his debtors could put his body in jail until the debt was paid. And Dick Jones did owe a debt, and they put his body in jail until the bill was paid—then he was brought home and buried right close to the house, near the front yard. I've seen the grave, but since then they took the grave up and carried it further away across a field."

"The man that built Bellvue, old man Dick Jones, I think he was."

Bellvue (also spelled Belle Vue and Bellevue) was built between 1804 and 1815 by Thomas Jones, son of General Joseph Jones and grandson of Roger Atkinson I. Jones purchased the property from the executors of John May in 1791. May, who seems to have grown up at Mayfield, had acquired, with various partners from the Petersburg area, hundreds of thousands of acres of land in Kentucky and nearby territories after the Revolution, but had been killed by Indians in 1790 on a trip to visit his lands. Thomas Jones, shortly after purchasing the Bellvue property, married Mary Lee, the daughter of Richard Lee of Lee Hall. According to Jeff O'Dell, Bellvue, upon completion, "may well have been the most elaborate house in the county." Thomas Jones's son, Richard Lee Jones, inherited the property in 1857 and died ca. 1870; was he the Dick Jones of Seay's story? Has anyone else heard of Dick Jones' story? (Photo courtesy of Jeff O'Dell from his book, Chesterfield County: Early Architecture and Historic Sites)

A TOUR GUIDE TO PETERSBURG'S CANAL

Refer to the fold-out map of the Upper Appomattox Canal, page 101.

The falls of the Appomattox in Petersburg divide the river into three parts: the Lower Appomattox, the falls themselves, and the Upper Appomattox. The *Lower Appomattox* is Petersburg's link with the sea, a tidal river from the wharves at Pocahontas Island down to the James. Sailing ships reached Petersburg through the *Lower Appomattox Canal*, a channel begun by the Lower Appomattox Company in 1802. Once dredged out again, this canal will revitalize Petersburg's harbor and bring new life to Petersburg's downtown.

Tidewater ends at the *Falls of the Appomattox*, five miles long, designated a State Scenic River by the General Assembly. It is popular with canoeists, who, armed with the *Appomattox River Canoe Guide* from the Chesterfield County Parks and Recreation Department, (804) 748-1623, can rediscover and explore the remains of the many dams, power canals, and mills which made Petersburg one of the most important industrial centers of the South.

Above the falls the Upper Appomattox Company began in 1795 the *Upper Appomattox Navigation*, designed for long, narrow wooden boats called *batteaux* which carried hogsheads of tobacco and the other products of the Appomattox valley down a hundred miles or more of winding river from above Farmville to Petersburg. The shallows were made navigable by channels called *sluices*, and by stone *wing dams* to direct the river into them. In the 1830's a series of dams with locks further tamed the waters of the upper Appomattox. Many of the works are now under Lake Chesdin, but the rest can still be explored by foot, canoe, and batteau.

The most impressive part of the Upper Appomattox Navigation was the *Upper Appomattox Canal*, a five-mile cut along the south side of the river, which carried batteaux from the head of the falls down to a canal basin in Petersburg. Completed by 1807, this canal was the company's major work and an engineering marvel of its time. It began at a dam across the river, wound high along the hillside, crossed several streams over stone culverts, descended a flight of four stone locks, crossed Indian Town (Rohoic) Creek on a huge stone-arch aqueduct and embankment, then ran into Petersburg to the canal basin. The canal served the people of the Appomattox River valley for a century, until navigation ended in 1902.

Today there is still much to see of the Upper Appomattox Canal. In *Ferndale Park* (renamed *Appomattox Riverside Park*), on Rte. 600 across the river from Matoaca, the canal still flows under the highway bridge. You can hike, canoe, or take a batteau two miles up the canal to the dam across the Appomattox. There was no towing path elsewhere along the river, but here along the canal two boatmen could get out on the towpath and walk a batteau, one man pushing on a pole attached at right angles to the bow, and the other at the stern. Note the spillways crossing the towpath—insurance against high water in the canal. When the canal was enlarged in the 1930's to carry more water down to a power plant, the stone culverts carrying creeks under the canal leaked so they were plugged up, forming ponds in the creek valleys—one of them is in Ferndale Park. At the *Abutment Dam* the canal entrance is blocked by a set of concrete control gates, but in canal days there was a stone *Guard Lock* here to let boats through. Some of the stonework is left, in the bank just upstream.

Downstream from Rte. 600, the canal is still watered for another mile, crossing a creek on a huge embankment to a spillway where it now ends. Someday these three miles of canal towpath, together with the embankments of two mill races (the Battersea and South canals) will make an exciting historic riverside trail all the way into Petersburg.

The rest of the canal has been filled in but part of it can be reached by car. From Ferndale Park, go south on Rte. 600 for 1.2 miles to the stop sign, left on East Rte. 226, and left at the traffic light onto North U.S. 1 for 0.4 miles to Rawlings Lane beside King's Barbecue. At the end of Rawlings Lane, accessible only on organized tours or by special permission from Virginia Power, are the remains of the flight of four stone locks called the *Toll Locks*, where tolls were charged, and the huge stone aqueduct over Indian Town (now called Rohoic) Creek.

Like Mayan ruins, the huge stone retaining walls of *Indian Town Creek Aqueduct* still stand despite a flood in Civil War times which destroyed the stone arch (a dam up the creek burst, washing it away) and later erosion which washed away the earth between the walls, leaving a deep canyon. A model of the site is in Petersburg's Siege Museum. One day, this historic site will surely be a park. Meanwhile, the walls are tumbling down and badly need stabilization.

Continue north on U.S. 1 for 1.1 mile to Battersea Lane, the first public street on the left past the Seaboard System Railroad overpass. Turn left, crossing *Commerce Street* and the tracks, then right at Battersea Plantation onto *Upper Appomattox Street*, named for the canal. The canal, now filled in, ran into Petersburg along the line of the railway tracks. Follow Upper Appomattox Street to the end, then right on North Dunlop Street to the top of the rise. The canal ran under you, into the Y on the left, formed by railway beds. The triangle formed by this Y and by South Street was once a square of warehouses surrounding the *Upper Appomattox Canal Basin*. To drive there, turn left on Commerce Street, left on South Street, and left before the next intersection, to the vacant lot behind the

Church of Christ.

For nearly a century this basin was a bustling center of commerce, handling great hogsheads of tobacco and the other products of the Appomattox valley's farms, forests and mines. Here the batteaux turned around and were poled back with barrels of whiskey and other city goods. There was never a navigable connection down to tidewater, so all the batteaux stopped here. The basin is now filled in, a time capsule from Petersburg's canal era which deserves protection and study. Any building excavation here should be carefully monitored. The watered part of the canal, the lock and aqueduct, and the canal basin deserve to be protected and placed on the National Register of Historic Places.

Return to South Street, turning left. Keep in the right lane and turn left onto *Canal Street*, following the Virginia State University sign. To the left, canal water tumbled down a spillway from the basin to the river, powering a series of mills, one after the other.

Bear left toward VSU at the bottom of the hill and park on the right before crossing the Ettrick Bridge over the Appomattox. Here near the foot of the falls you can explore two of the river's many other mill races. These used the river's water power to run factories and mills, but were not navigated by boats. On the Petersburg side, one can walk 0.3 mile up the *South Canal* to a dam, and also explore the mill foundations just downstream from the bridge. The *North Canal* on the Chesterfield County side is now the centerpiece of *Ettrick Riverside Park*. This canal has been restored to operate a small hydro plant and eventually a working mill. A trail leads up to the dam. This is an ideal place to admire the beauty, power and danger of the falls and to see for yourself why the boatmen so desperately needed the *Upper Appomattox Canal*!

FOR FURTHER READING

Appomattox River:

"Appomattox River Canoe Guide," to the Appomattox State Scenic River, free flyer from the Chesterfield County Parks and Recreation Department, P.O. Box 40, Chesterfield, VA 23832, (804) 748-1623.

Carter, E. C., ed., *The Virginia Journals of Benjamin Henry Latrobe,* Yale University Press, 1977; see also *Virginia Cavalcade,* Spring 1959 and Summer 1979.

Couty, John, "Map and Survey of the Appomattox River from the Town of Petersburg to Planterstown," 1834. The only map ever made of the Upper Appomattox Navigation. Copies available from the City Engineer's Office in Petersburg, Map # B-11-72.

Cocks, Edmond D., "The Appomattox," *Virginia Geographer,* XVI (1984) 57-67.

Henderson, William D., "Rapids and Power: The Appomattox River and Electrical Power in Petersburg, Virginia," *Virginia Cavalcade,* Spring 1978.

Jones, Richard L., "Appomattox Frontier," unpublished manuscript, Petersburg Public Library.

Trout, W. E., III, *The Appomattox River Atlas,* historic sites on the Appomattox and its navigable branches, 50 pp., 1990. $5 ppd. to "Petersburg Department of Tourism," from Centre Hill Mansion, Petersburg, VA 23803. Trout is the author of a series of other Virginia river atlases, for the James River, Rivanna River, Maury River, Rappahannock River, and Goose Creek. Others are in the works, including *The Falls of the Appomattox Atlas.* For a price list, write to Richard A. Davis, V.C.&N.S. Sales, Rte. 2, Box 254, Lexington, VA 24450.

"Upper Appomattox Canal," a map and tour guide of the Upper Appomattox Canal by W. E. Trout, III. Free flier available from the Chesterfield County Parks and Recreation Department. Reproduced on pages 85-87 and 101 of the present volume.

For Further Reading

Batteaux, Canal Boats, and Canals:

American Canals, illustrated quarterly of the American Canal Society, 117 Main Street, Freemansburg, PA 18017.

Kirkwood, James J., *Waterway to the West*, an illustrated history of the James River and Kanawha Canal, National Parks & Conservation Association, 1963. Available from the National Park Service or V.C.&N.S. Sales.

Terrell, Bruce G., "The James River Batteau," *Virginia Cavalcade*, Spring 1989, pp. 180-191. For more details see his Master's thesis on the subject from East Carolina University. Soon to be published.

The Tiller, illustrated quarterly of the Virginia Canals and Navigations Society, c/o The Alexandria Waterfront Museum, 44 Canal Center, Alexandria, VA 22314.

Trout, W. E., III, The American Canal Guide, Part 5, a guide to the historic canal resources of Delaware, Maryland, and Virginia, American Canal Society, 1992. Available from Keith Kroon, A.C.S. Sales, 2240 Ridgeway Ave., Rochester, NY 14626.

The Virginia Batteau Journal, publication of the James River Batteau Festival, Inc., c/o Sue Pechman, 33 Moorman Rd., Madison Heights, VA 24572.

Other publications of the American and Virginia canal societies. Ask for publications list.

County and City Histories:

Bailey, James H., et al, *Old Petersburg*, Richmond, Va., Hale Publishing, 1976.

_____, *Pictures of the Past: Petersburg Seen by the Simpsons, 1819-1895*, Petersburg, Fort Henry Branch, Association for the Preservation of Virginia Antiquities, 1989.

Henderson, William D., *Gilded Age City: Politics, Life and Labor in Petersburg, Virginia, 1874-1889*, Lanham, Md., University Press of America, 1980.

_____, *The Unredeemed City: Reconstruction in Petersburg, Virginia, 1865-1874*, Washington, D.C., University Press of America, 1977.

Jones, Richard L., *Dinwiddie County: Carrefour of the Commonwealth*, Whittet and Shepperson, Richmond, 1976.

Lutz, F. E., *Chesterfield, an old Virginia County*, Byrd Press, Richmond, 1954.

O'Dell, Jeffrey M., *Chesterfield County: Early Architecture and Historic Sites*, Chesterfield County, 1983.

Scott, James G., and Edward A. Wyatt, IV, *Petersburg's Story: A History*, Petersburg, Titmus Optical Company, 1960.

Other:

Davis, Charles Hall, compiler, Virginia Consolidated Milling Company scrapbook, December 1905, used courtesy of the Petersburg Public Library, where it is located.

McKenney, Carlton N., *Rails in Richmond* (and Petersburg), Interurban Press, Glendale, CA, 1968.

Rulau, Russell, *Hard Times Tokens*, Krause Publications, 1987.

Wyatt, Edward A, IV, "Rise of Industry in Ante-Bellum Petersburg," *William & Mary Quarterly*, 2nd Series, 17 (1937).

For More Detailed Research:

Annual Reports of the Upper Appomattox Company and other records in the Virginia State Library.

Other reference material in the Petersburg Public Library, Virginia Historical Society, Chesterfield Historical Society, Chesterfield County Public Library, other local libraries, and county courthouses.

INDEX

Abby Aldrich Rockefeller Folk Art Center, 14
Abutment Dam, 56, 57, 58, 86, 101
Adams, Woody, 66
Allen's Marina, 36
American Canal Society, xiii, 90
Andrews, Hugh Allen, xvii
Andrews, Mr., 44, 45
Andrews, Peter, viii
Appomattox River, 89
    falls of, xiii, 85
Appomattox River Water Authority, 56, 60, 62
Appomattox Riverside Park, see Ferndale Park
Aqueduct, see Indian Town Creek Aqueduct
Archer, Woody, 66
Astoria Hotel, 45
Atkinson, Robert, 56
Atkinson, Roger I, 72
    Roger II, 72
Atkinson, Thomas, 73
Aunt Harriet's furniture, 18, 19
Ayers, Joe, xviii

Bailey, Dr. James H., xviii, 90
Basin, canal, see canal basin
Basin Mill, 22, 23
Batteau, xvii, 14-21, 30, 90
    cargo, 3, 15, 18, 77
    cover, ii
    dimensions, 5
    operation, 1, 3, 85, 86
    see also ferry boat; poling
Batteau Day, 10
Battersea, xiv, 86
    Canal, 86
Beadle's Dummy, 76, 77
Beadle, George, 76
Beech Island, 70
Belcher, Syl, 75
Bell, Fred R., ii, xviii, 30, 39, 52
Bellvue, 82, 83, 101
Bevil's Bridge, 3, 19
Black Billy, 78
Black's Canal, 20, 38, 101
    Wall, 17, 21
    Dam, 49
Blue Ridge Cotton Factory, 22
Boisseau, Holmes, Jr., 40, 41
Bosher, Mr., 21
Bowling alley, Ferndale Park, 53, 55
Bowman, Oscar, 19
Boydton Plank Road, 25
Brady, T. T., 26, 42
Brasfield, George, 62, 63
Brasfield Dam, see Lake Chesdin Dam
Bridges, 18, 19
Broad Rock Falls, 17
Brown, W. M., 65

94 Index

Browning, Lyle, 4
Bruce, David, 26
Bryant, Hootie, xvii
Bunting, Mr., 49

Cabin, Seay's, 46, 47, 48
Campbell's (Ettrick) Bridge, 22, 81
Canal basin, Petersburg, xvii, 3, 19, 22, 23, 85-87, 101
Canal basin, Richmond, xviii, 4, 10
Canal Street, 77, 87, 101
Canals, see Battersea, Black's, Caudle's, Chesapeake & Ohio, Kanawha, Lower Appomattox, North, Petersburg's, South, Upper Appomattox
Canoe, log, 7, 42, 43
Carousel, Ferndale Park, see hobby horses
Carter, E. C., 89
Carter, Randy, 21
Caudle's Canal, 8, 17, 101
  Lock, 12, 13
Central Lunatic Asylum (now Central State Hospital), 31, 76, 77, 101
Chapel Street, 77
Chesapeake & Ohio Canal, 3
Chesterfield County Museum, 7, 42
Chesterfield County Parks and Recreation Department, 85, 89
Chesterfield County Public Library, 91
Chesterfield Historical Society, 2, 10, 43, 91
Christ and Grace Episcopal Church, 40, 41
Civil War (War of Northern Aggression), 24, 25, 86
Clement Town, 18, 20, 34
Clements, Frank, 31, 32
Cocks, Edmond D., 89
Cohoon Pond, 75
Coin, NOT ONE CENT, 82
Colonial Williamsburg Foundation, xviii, 14
Commerce Street, 23, 86, 101
Concrete work, 5, 35, 71
Confederate Powder Mill, 28

Couty, John, 19, 24, 28, 89
Cox, T. D., xvii
Crowder, Willie, xvii

Dalton, Gary, xviii, 1
Dams, see Abutment Dam, Black's, Goode's, Gould's, Lake Chesdin, Lake Margaret, Lee's, Matoaca Mill, wing dams
Dance hall, Ferndale Park, 53
Davis, Charles Hall, see Virginia Consolidated Milling Company
Deep Creek, 3, 49, 99, 101
Dog (little), 51
Dooner, Hugh, 78, 80
Dunlop Street, 23, 86
Dunnavant, Mrs. H. L., xviii, 54
Dynamite, 57, 77

Ettrick Bridge, see Campbell's Bridge
Ettrick Riverside Park, 87
Exley, John, viii

Farmville, 3, 7, 11, 14, 18, 19, 22, 40, 85, 101
Ferndale Park, ii, xvii, 10, 30, 31, 50-59, 71, 76, 86
  streetcar, 50, 52, 76
Ferry boat, vi, xvii, 5, 30-33, 51, 52; cable, 31, 32
Fishing, 1, 34, 36, 69

Gold (lost), 20
Goode's (Gould's) Bridge, 45
  Falls and dam, 37, 101
Gould, Frank Jay, 44, 45, 49
  dam, 44, 45
  mansion, 44, 45
Graham, William J., xviii
Granite Grove, 77
Grosfils, Catherine H., xviii

Hand, Orval, 62, 63
Hardy, Tom, 21
Harrison, George, Jr., 40, 41
Harrison, Mary, 40
Hartman, Richard D., xviii, 62
Haxall, Theodore W., xviii, 65

Index   95

Henderson, William D., 89, 90
Hickory Road, 78
High Bridge, 14
High Street, 22, 23, 77, 81
Hillier, Dr. Joseph C., xviii
Historic Petersburg Foundation, xiii-xv
Hobby horses, Ferndale Park, 53, 54, 55
Holt, Alpheus, xvii
Holt, Hubert, xvii, xviii, 36, 71
Holt, Mary Louise, 36
Holt, Tom, xvii
Houseboat, 7, 44-49
Howlett, 67, 78
    Mill, 68, 69
Hunter, Peter George, viii

Ice cream parlor, Ferndale Park, 53, 56
Ice jam, 37
Indian Town Creek, 24, 101
    Aqueduct, 5, 24-28, 85, 86, 101
Indian Town Mill, 28
Indians, 7, 73, 82, 83
Iron Mountain, Michigan, 66

James River Batteau Festival, xviii, 90
Jamestown Exposition, 33
Jones, Dick, 83
Jones, Dr. Bolling, 78
Jones, General Joseph, 83
Jones, Richard L., xviii, 89, 90
Jones, Richard Lee, 83
Jones, Thomas, 83

Kanawha (James River & Kanawha) Canal, 3, 90, 99
Kirkwood, James J., 90
Kofron, Mr., 74

Lady Caroline, Ballad of, 72
Lafayette's Rock, 64, 65, 66
Lake Chesdin, 1, 20, 21, 42, 46, 47, 48, 60-63, 85
Lake Chesdin (Brasfield) Dam, 8, 12, 60, 62
Lake Margaret Dam, xviii, 67, 68

Lassiter, Daniel W., 77
Lassiter's Silk Works, 28
Latrobe, Benjamin Henry, 11, 34, 89
Lee, Mary, 83
    Richard, 83
Lee Hall, 83
Lee's Dam, 24, 25
Leonard Hardware, 32
Level, running a, 46
Locks, 5
    mud sill, 21
    see also Toll Locks
Lord Chesterfield Batteau, 10, 11
Lost art rediscovered, 74
Louisville, Kentucky (rabbit from), xvii
Low Wall Sluice, 8
Lower Appomattox Canal, 85, 99, 101
Lutz, F. E., 90

Mansfield Plantation, 72
Marick, Mr., 21
Markel, Art, 10
Martin, William J., xviii
Masonic Lodge (Blandford No. 3), 65
Matoaca, ii, xvii, 1, 5, 6, 86, 101
Matoaca Mill, xvii, 18, 30, 31, 78, 79
    Dam, 70
    office building, 80
    picker house, 12, 13, 79, 80
May, John, 83
Mayfield, 83
McCarter, M. D., 50
McKenney, Carlton N., xviii, 50, 54, 76, 91
Meadows, Mrs. Rosa, 9
    Mrs. Virginia, 9
Miller, Lewis, 14
Mills, 89
    Petersburg Basin, 22, 23
    see also Confederate Powder Mill, Indian Town Mill, Lassiter's, Matoaca Mill, Seward & Munt's Grist Mill
Morton, S. D., ii, iii
Mule power, 5
Munt, H. L., 81
    Mill, 22, 81

Namozine Creek, 36
New Hampshire (mysterious man from), 64, 65, 66
New River, 16, 20
Norfolk, 75
Norfolk & Western (Norfolk Southern) Railroad, 3, 22, 24, 26, 27, 51, 75
Norlington, Pat, 52
North Canal, 87
Nottoway River, 5

Oakdale Avenue, 77
O'Dell, Jeffrey M., xviii, 6, 72, 80, 83, 90
Old Virginia trunk factory, 35
Olger's Store, 37
Olive Hill, 72, 73, 74, 101
Paddle, Seay's, 37

Palms Ice Cream Company, 31, 32
Partin, Danny Rubin, vii
Pease, Fred, 43
Perpetual motion, 33
Petersburg, siege of, 24
Petersburg & Asylum Railway, 76
Petersburg Iron Works, 5
Petersburg Public Library, 22, 79, 81, 89, 91
Petersburg's (Upper Appomattox) Canal, tour, 85-87
Phillips, Joe, 77
Picker House, 12, 13, 79, 80
    see also Matoaca Mill
Pitts, C. R. (Dick), Jr., vi, xiii, xviii
Planterstown, 3, 101
Pocahontas Basin, 39
Pocahontas Island, 85, 101
Poindexter, Mr., 49
Poling, 3, 14, 15, 20
    poles, 15-17
    pole tips, xviii, 7, 16, 17
Pollock, Edward, 81
Poole, John W., mill, 22
Poplar Lawn Park, 39
Potholes, 39
Purcell, Warren C., xviii

Quakers, 72

Randolph Plantation, 65
Rappahannock Navigation, 21
Rawlings Lane, 86
Regan, Bosh, 78, 79
Reynolds Metals Company, viii, xviii
Richmond Public Library, 9
Ritchings, Arthur W., xiii, xiv
River Road, 7
Rock Castle, see Gould's Mansion
Rohoic Creek, see Indian Town Creek
Rohoic Dam, see Lee's Dam
Roller Mill, 22
Roper, Philip R., Jr., xviii, 40, 67, 68, 69
Roper Brothers Lumber Company, 67
Rose, Col. Llewellyn, 62, 63
Rosenstock, Anthony, home of, 6, 7
Rowlett, see Howlett
Rucker-Rosenstock department store, 7
Rulau, Russell, xviii, 82, 91
Russell, A. J., 25, 26

Scott, James G., 90
Seay, Frank, xviii, 9, 38, 41, 45, 55, 66, 70
Seay, Gordon E., 9
Seay, James Malcolm (Mac), Sr., xviii, 9, 67
Seay, James M., xviii
Seay, James Washington (Jim),
    batteau, 15
    biography, 1-9
    blasting Lafayette's Rock, 64-66
    building dams, 67-70
    cabin, 46-48
    dugout canoe, 42, 43
    ferry boat, 30-33
    fishing with Mary Louise, 36
    home of, 6, 7
    houseboat, 7, 44-49
    Lake Chesdin voyage, 63
    Olive Hill, 72-74
    NOT ONE CENT, 82
    watch, vi
    see also other stories
Seven Pines, Battle of, 67

Index

Seward, Simon, 81
Seward & Munt's Grist Mill and Trunk Factory, 81
Shakespeare, William, 66
Sheppard, W. L., 20
Shooting gallery, Ferndale Park, 53, 55
Siege Museum, 86
Slaves, 73, 82
Sluices, 85
Smith, Blackford, xvii
Snakes, Jim Seay's, 69
South Canal, 86, 87
South Street, 22, 23, 86
Southside Railroad, see Norfolk & Western
Steam drills, 58, 59
Steering oar, viii, xviii, 3, 7, 10, 14, 16, 17, 20, 33
Sterns, John N., mill, 22
Streetcars in Ferndale Park, 31, 50-52, 54, 76
Suffolk, 75
Superintendent's Residence, Ferndale Park, 56
Sutherland, 37
Sweep, see steering oar
Swingers, Ferndale Park, 51
Sycamore Island, 56, 57

Tape recorders, v
Taylor, Olin L., 43
Terrell, Bruce G., 90
Theater, Ferndale Park, 52, 53, 54
Toll Locks, 3, 5, 28, 29, 86, 101
Traylor, Mr. & Mrs. Rubin, vii, xviii, 78
Treasure (lost), 20
Trout, Mrs. W. E., Jr., 12, 13
Tunnel of Love, Ferndale Park, 53

Upper Appomattox Canal, 3, 21, 54, 57, 85
Upper Appomattox Company, ii, 3, 11, 85
 Roger Atkinson II, 72
 stock certificate, 9
Upper Appomattox Navigation, xiii, 3, 85, 99, 101

Upper Appomattox Street, 86, 101

Valentine Museum, 65
Van Deventer, Fred, 2
Vaughan, Mrs. Audrey, 9
VEPCO, see Virginia Power
Victrola, Ferndale Park, 53
Virginia Canals & Navigations Society, xiii, xv, 89, 90
Virginia Consolidated Milling Company, 22, 23, 79, 91
Virginia Department of Historic Resources, 17
Virginia Historical Society, 91
Virginia Passenger & Power Company, see Virginia Power
Virginia Power, 28, 44, 51, 56, 58, 59, 71, 75, 86
Virginia State University, 65, 87
Virginia State Library and Archives, 11, 19, 34, 91

Wading pool and fountain, Ferndale Park, 52, 53
Wallace, Lee A., Jr., ii, 30
Walls, Mr., 74
Ward, Dulaney, xiii, xiv, xviii
Washington, George, xv
Watches, vi
Webber's stone, 38
Wells, C. Clagett, 65
Wessells, Tom, xvi
Whiskey, 19, 34
Wilburn, Willie, 77
Wilcox Lake, 40
Wing dams, 85
Woodpecker Road, 67
Woodruff, Mrs. Gretchen, xiii, xviii
Wyatt, David, 21
Wyatt, Edward A., IV, 90, 91

The Upper Appomattox Navigation

←21 miles to Planterstown, head of navigation

- Royaltown Mills L&D
- Tucker's Ford L&D
- Brackett's Bends L&D
- Genito Mills D,C,2L
- Randolph's Gravel L&D
- Clement Town Mills D,C,2L
- Wood's Ford L&D
- Stony Point Mills D,C,2L
- Flat Creek Navigation
- Holcomb's Mill D,C,L on Deep Creek Navigation
- Wing dams & sluices
- Webber's Ford L&D
- Chesdin (Brasfield) Dam
- Eppes' Falls L&D
- Gould's L&D
- Morton's Mill D,C,L
- Buffalo Creek
- Prince Edward Mill Dam
- Routn's Sluices L&D
- Jamestown L&D
- Bush River L&D
- Union Mills D,C,2L
- Exeter Mills L&D
- Black's Canal
- Caudle's Canal
- Lower Appomattox Navigation
- Upper Appomattox Canal

Covered by Lake Chesdin

FARMVILLE — PETERSBURG

0 ——— 10 miles
D = Dam   C = Canal   L = Lock

The Upper Appomattox Canal
(The arrows are for use with the self-guided tour on pages 85-87.)

— watered
▬ ▬ filled in
• • • • mill races

Chesterfield County

- Abutment Dam, head of canal and the falls
- Olive Hill
- The Falls
- Ferndale Park (Appomattox Riverside Park)
- Matoaca
- Bellvue
- Spillway
- Toll locks and aqueduct
- Ettrick
- Canal basin
- Pocahontas
- The Upper Appomattox Canal
- Upper Appomattox St.
- Commerce St.
- Washington St.
- PETERSBURG
- Rohoic Creek

Dinwiddie County

SCALE 0 — 1/2 — 1 mile

Adapted from the Dinwiddie County map, Virginia Department of Highways and Transportation

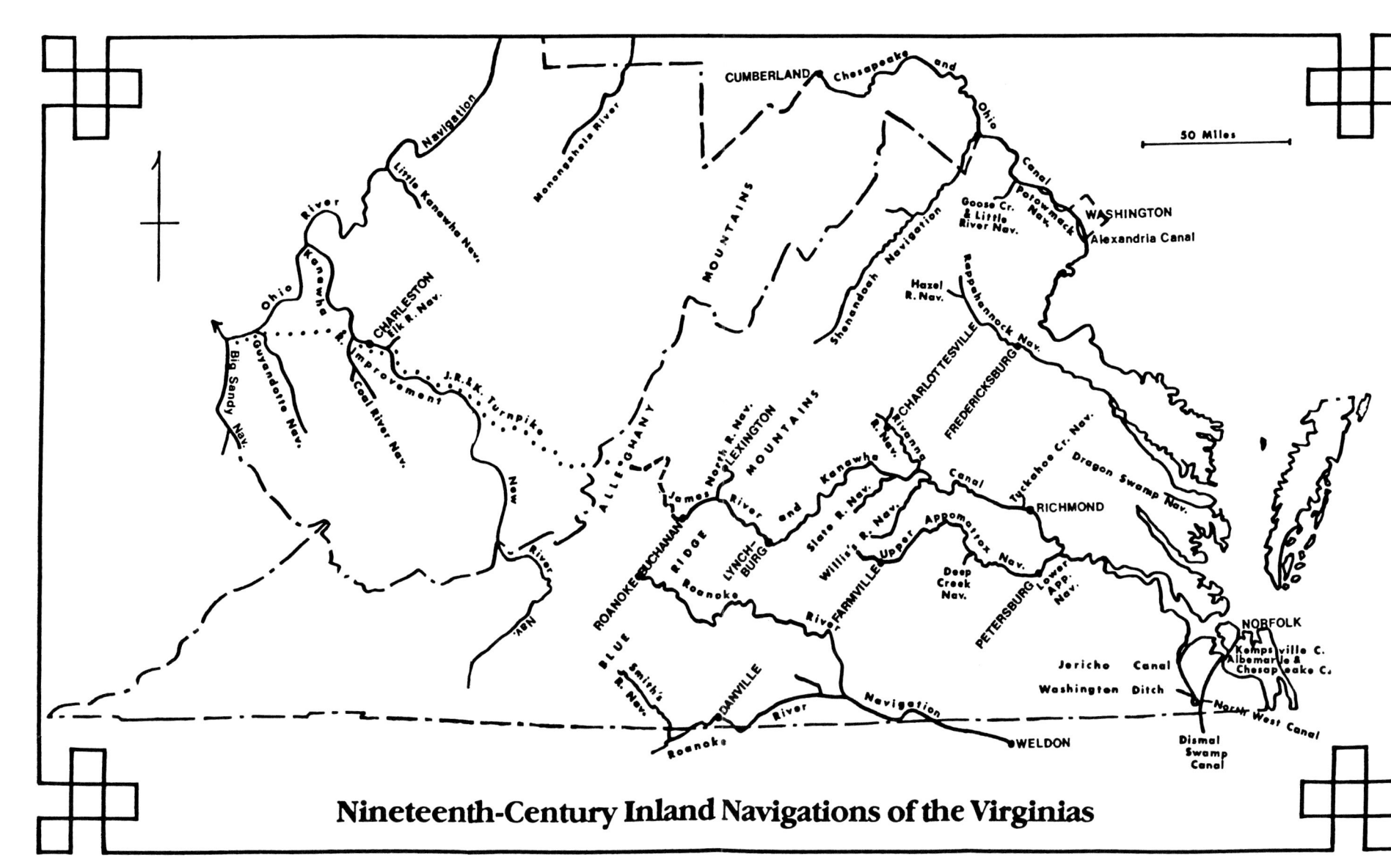
Nineteenth-Century Inland Navigations of the Virginias